Patrick,

Always Great Working with You and Your MT 2000! Keep Milling,

Dan M.

PAVING THE WAY
ASPHALT IN AMERICA
DAN McNICHOL

10.12.06

NATIONAL ASPHALT PAVEMENT ASSOCIATION

PAVING
THE WAY

ASPHALT
IN AMERICA

DAN McNICHOL

PAVING THE WAY: ASPHALT IN AMERICA

For information contact:
National Asphalt Pavement Association, 5100 Forbes Blvd., Lanham, MD 20706
888-468-6499 (toll-free)
www.hotmix.org

Publisher:	Margaret Blain Cervarich
	National Asphalt Pavement Association
Editor:	Jane F. Neighbors
Design:	Richard J. Berenson
	Berenson Design & Books, LLC, New York, NY
Production:	Della R. Mancuso
	Mancuso Associates, Inc., North Salem, NY

ISBN 0-914313-04-5

Printed in Canada

First Printing

Contents

Philadelphia, July 12ᵗʰ 1864

Mr Williamson Dr I S Williams

To DANIEL McNICHOL & BRO. Dr.

CONTRACTORS FOR CURB STONE SETTING & PEBBLE PAVING,

No. 120 JUNIPER ST., ABOVE ARCH,

ALL ORDERS PROMPTLY ATTENDED TO.

Curb Reset No 23 Rye st
at @ 10¢ pr ft   1 60
Regulation @ 2¢ pr ft   32

$1.92

Payment

Daniel McNichol & Bro

"Sunny Jim" McNichol and an invoice from
his father's firm, Daniel McNichol & Brothers

Dedication

JAMES "SUNNY JIM" McNICHOL was born in Philadelphia, Pennsylvania, on July 3, 1864. His father and uncles, Irish immigrants, contracted for small paving and curb-repair jobs under the name Daniel McNichol & Brothers. Branching out, Sunny Jim—named for his bright outlook on life—joined his brother Daniel. "Making the plunge," he and his brother borrowed $1000 and purchased two carts and three horses. The brothers made a name for themselves converting the City of Brotherly Love's pavements from Belgium blocks to asphalt.

By 1909, their McNichol Paving & Construction firm was one of the largest in the country, employing 7000 men. The firm built and paved the notable avenues Roosevelt Boulevard and Benjamin Franklin Parkway and were contractors on major public works projects such as the Philadelphia water-filtration plant and New York City subways.

Sunny Jim was elected to Philadelphia's Select Council as a ward leader in 1898, where it was said that "Its splendid streets are Philadelphia's crowning glory." In 1904 he was elected to the Pennsylvania State Senate, where he served until his death in 1917. His passing was front-page news, with the *Philadelphia Inquirer* reporting that his was "one of the largest funerals ever held in Philadelphia." Some 300 policemen guided 10,000 mourners to the city's cathedral. A social record of the time gushed, "The city of Philadelphia has probably the best paved streets in the world and the great work contributed to in by far the greatest degree by Senator McNichol."

Foreword

By Mike Acott, President, National Asphalt Pavement Association

THERE'S SOMETHING about asphalt people: They love asphalt. They like the smell of it, they even like getting it on their shoes when they visit a paving site. They never get tired of looking at pavements and comparing notes. Some of them say, "If you cut me, my blood runs black." Take an asphalt group to Rome, and all they will all talk about is how the ancient city's cobblestones are finally being replaced with asphalt.

The story of asphalt in America is the story of the people of asphalt. In the very early days, more than 100 years ago, starting a road-building business did not take a lot of capital, just a lot of hard work—making it a good business for immigrants and others who had plenty of energy and ambition but not much money. Italians and Irish in the Northeast, Germans and Scandinavians in the Midwest, English in the Mid-Atlantic States and the Deep South—these and other national and ethnic groups have contributed to the asphalt industry. Many of these firms still exist today, still carrying the name of the original owner.

These early businesses were family operations, and today many a company in the industry is operated by the second, third, or fourth generation of the founding family. It was common for the children of the owners of asphalt paving firms to learn the ropes of the family business the hard way—some said the only way—by running a roller, driving a truck, or pushing a rake. Often the children of employees did the same by following in their parents' footsteps, creating a multigenerational legacy that brings real meaning to the term family business.

Farmers were among the first asphalt contractors. Often, a farmer would find gravel on his land that was suitable for putting a hard surface on the farm's roads. His tools would be a grader and a roller. From putting gravel on the surface to mixing it with asphalt was a small jump. He might perform the same service for his neighbors, and soon work himself into the position of being the local road contractor.

Many of today's asphalt firms came out of the aggregate industry. Once a quarry or sand-and-gravel operation was in place, putting an asphalt plant next to the source of aggregate made a lot of sense.

Asphalt producers and contractors tend to be imaginative, self-reliant people who are always looking for a better way to do things. Many companies were

Andrew Oneglia, an immigrant to America from Italy by way of Argentina, in the driver's seat of his three-wheeled steamroller in the early 1930s. He and a partner founded O & G Industries, now one of Connecticut's largest contractors.

founded by the innovators of their times, the people who developed a particular process or mix type. One of those was Joseph Hay Amies, a doctor of divinity and Lutheran clergyman who started patenting his own materials for paving in 1892. His Amiesite mixtures were used in many states, and today's IA Construction Corporation was first known as Interstate Amiesite. The tradition of innovation by asphalt contractors—not just equipment manufacturers—continues today.

Veterans returning from abroad have formed a sizeable group of workers entering the asphalt industry. The ability to move troops and matériel played a major part in America's victories in two World Wars. Whether they gained experience with road building in the mud of France or runway paving in the Pacific Theater of War, veterans came home ready to build roads that would further the peaceful expansion of America's culture and economy.

In the great migration of America's population from the countryside to the cities in the 20th century, farm workers were often attracted to road building. They liked the outdoors life, and they already knew how to drive heavy equipment and keep it operating. Some started their own companies, some took jobs working for others; either way, the industry offered opportunities that were a perfect fit.

Carey Moore with her father, Jim Madison, and older brother, Jimmy, in front of their Memphis home.

While asphalt, like other contracting businesses, was for a long time a mostly male preserve, it was also part of women's lives. Carey Moore, recalling her father's long hours during a postwar Tennessee childhood as he built up Lehman-Roberts Company, says, "I am an asphalt legacy. Grandfather, father, brother, husband, son-in-law, have been building America's roads for almost 100 years. It is all I have ever known, as the men in our family have left for work before the sun was up and returned long after it set.

"We waited dinner for Dad, even when his homecoming came just before our bedtime. I can remember the sounds of his heavy, rock-laden shoes hitting the kitchen floor as he entered in his sock feet. I recall the lingering asphalt smells that clung to his clothes. The roadside dust gave a beige coating to his hair and complexion. We listened attentively to the recounting of his day and the tons of asphalt laid and the deadlines and the night jobs and the broken-down equipment. There was little we understood, but we were thankful to have Dad include us in the daily briefing."

George Wilson was a second-generation contractor. His father, Gaines P., had started his career as an employee of the R. B. Tyler Company in Louisville, Kentucky. Tyler was a large multistate contractor, one of whose innovative projects was paving the banks of the upper Mississippi River in the late 1940s. George

A tug moves huge storage tanks up the Cumberland River to their new location at an asphalt terminal in Kuttawa, Kentucky. Each tank was 48 feet in diameter. The terminal was built in 1954 for Peter Wilson's grandfather's asphalt business.

had relocated to Louisiana and was ready to start his own asphalt company in 1972, when he asked his wife what to name the company. Peter Wilson, his son, recalled later, "Dad picked up the phone and called Mom, and said, 'What are we going to call this business?' She said, 'Since the office is on Barriere Road, why not call it Barriere Construction?'"

The naming was not Betty Wilson's last contribution to the company. Her husband died just a few years later, and she became chairman of the business. Under her leadership, Barriere Construction grew into one of the largest asphalt contractors in Louisiana. When she retired fifteen years later, she passed responsibility to her sons, George, Peter, and Bert.

One of the first large companies in asphalt was the Warren Brothers Company, based in Boston. The company started its life in 1900 in competition with Amzi Lorenzo Barber, the Asphalt Tycoon. Barber's asphalt empire, built on an attempt to corner the market on asphalt, eventually failed, but the Warren family's business proved more resilient. The brothers were pioneers in every aspect of the industry, from refining crude oil to designing pavement mixes to marketing equipment for both plants and paving operations. As a result, it thrived for decades, opening its 100th plant in 1965. The company merged with Ashland Oil and Refining in 1966 and became Ashland Paving and Construction in 1982. Under the name APAC, Inc., the company is still one of the largest hot-mix asphalt producers in the country.

The industry today has many multistate and multinational companies that operate scores or even hundreds of asphalt plants. Small or large, asphalt companies share a dedication and commitment to quality and innovation.

In asphalt, the work environment can be challenging. Both manufacturing and construction take place in the great outdoors, with much of the work performed in full view of the customer. Often the elements are a factor; often cars and trucks are whizzing by a few feet away. The paver and roller operators, shovelers, rakers, paving superintendents, and plant managers are on the front lines. They are backed by an army of administrative and support personnel. All of them together are the unsung heroes of the open road. This book honors them, their work, and their commitment.

These people are the strength of asphalt in America.

CHAPTER ONE

American Asphalt

"As new and greater road-systems are added year by year they are more splendidly built. I foresee that road will soon be architecture too ... great architecture."

—Frank Lloyd Wright

AMERICA'S VEINS AND ARTERIES run black with asphalt. Every hour of each day and every day of each year, asphalt roads are pumping life into cities, out of the extremities of the country's hinterlands, and back again. Life as we know it is largely due to our unmatched road net. In the United States, nearly every mile of our paved roads is surfaced with asphalt.

Our roads have helped make America the most powerful nation on earth. As a world power, the United States has influence and wealth greater than that of the Roman Empire at its height. It's no coincidence that the two share a common physical trait—the greatest road systems of their times. Both Rome and the United States came to power over the road—Rome over its stone roads and America over its asphalt roads.

Americans travel and trade over their roads in a way that the civilized world has never before witnessed. So dependable are the millions of miles of asphalt roads that manufacturers can schedule shipments of materials to arrive within 15 minutes of their being needed for assembly-line production. The process is called just-in-time-inventorying, and it gives the United States a big advantage in competing against low

Signs reading "asphalt pavement only" are not posted in the National Parks, but they might as well be.

wages abroad. Speaking of working: Every day, Americans commute to locales that were once beyond their reach, using the same asphalt roadways that make the firms competitive.

The deployment of armed forces overseas begins at home. In times of conflict, troops, weapons. and munitions are sped from the U.S. Army's forts to the U.S. Navy's ports in an exercise called forts-to-ports. Because of the roads between the military's vital links, the U.S. forces are able to perform this type of operation faster than any military in history. Three quarters of shipments made inside the nation's borders by the Department of Defense are made by truck over asphalt roads.

Some people assume that asphalt is not environmentally friendly. In truth, it is the most recycled material in the nation. More asphalt is recycled than paper, glass, metal, or wood. More than 70 million tons of asphalt pavement that is reclaimed from roadways every year is reused or recycled. If all the asphalt pavement that is reclaimed every year were piled up, you would see a mountain of recycled roadway 749 feet high and 1398 feet in diameter.

Since the 1970s, the asphalt industry has made strenuous efforts to reduce emissions at its facilities. In cooperation with the Environmental Protection Agency (EPA), the hot-mix asphalt industry established tests and standards to judge its impact on the environment. As a result, in 2002, the EPA deleted asphalt from its list of major sources of hazardous air pollutants, making it one of the few industries to have been delisted.

Signs reading "asphalt pavement only" are not posted in the National Parks, but they might as well be. Dust, mud, and erosion are the enemies of our treasured recreational areas. Dust kicked up by passing vehicles spreads over plants and wildlife, often killing the plants and altering the ecosystem. Asphalt is an ally in the National Park Service's stated quest to have any development "lie lightly on the land." It is the pavement you'll find when driving on more than 5000 of

EXITS 14-14A-14B-14C 1 MILE
78   1   22
Newark Airport
Holland Tunnel

Americans practice their faith, educate their young, care for their elderly, and seek business over asphalt roads.

the 5300 miles of paved roads in the National Parks. It is also used extensively in the National Parks for biking and hiking trails that protect vegetation from being trampled.

The United States of America is a road society—a nation relying on its roads. We pride ourselves on moving freely across one of the largest land masses in the world. Americans practice their faith, educate their young, care for their elderly, and seek business over roads. From our driveways and parking lots to our local streets to the two-lane highways and 16-lane superhighways, our lives are as we know them because of asphalt.

A hundred years ago, Americans sang songs and wrote poetry about asphalt, celebrating the improvement it was making in their lives. Today, asphalt roads are so ubiquitous as to be invisible. Reverse engineering the tale of asphalt can illuminate the real story of America's development.

LOOKING BLACK

Paris, London, and Berlin were the first capitals to boast of having asphalt avenues. The world marveled at the smoothness, quickness, and quietness with which carriages were moving across the new mystical black carpets.

Before asphalt, dust, mud, or worst of all, disease resulted from the dirt, cob-

Like a long distance runner in an endless test of endurance, asphalt has outlasted and outperformed all other road pavements.

blestone, brick, and wooden-block roads. Manure and other refuse, trapped in nooks and crannies, was blamed for epidemics. Because asphalt was smooth, it was easier to clean; and because asphalt repelled water, it was inhospitable to bacteria. Americans in crowded, noisy, and dirty cities began demanding the cleanliness of asphalt streets.

The United States celebrated its Centennial by asphalting its most prestigious boulevard. President Ulysses S. Grant decided that the nation's most notable street, Pennsylvania Avenue in Washington, D.C., should be paved with the finest, most durable material available. At the time, Washington's muddy avenues and rotting wooden streets—that's correct, wooden—made for an international embarrassment. Politicians began debating and planning a relocation of the nation's seat of government—anywhere but in muddy Washington.

The national capital's redemption came with its asphalt pavements. In 1876, Pennsylvania Avenue received the first successful application of asphalt pavement on a street anywhere in the United States. Investing heavily in the unproven material, Washington began paving the city's now famous avenues including Pennsylvania Avenue, Constitution Avenue, and Massachusetts Avenue.

The U.S. capital and its avenues became a major attraction. By the end of the 1800s, tourists were flocking to Washington, not only to see the White House and monuments but also to see its famous asphalted avenues.

Asphalted avenues were joined by paved side streets and eventually highways, but it took decades to get there. Mobilizing during World War I helped Americans see the need for durable roads that could withstand rain, snow, heat, and most important, the beating of heavy traffic—both in tonnage and in volume.

Asphalt had many challengers as the pavement of choice: brick, stone, concrete, wood, and steel. But like a long distance runner in an endless test of endurance, asphalt has outlasted and outperformed all other road pavements.

ROCKS AND ROLLING

America's asphalt roads are really roads of rock. Asphalt, making up about five percent of the road's material, is the magic that makes the road black, smooth, watertight, and durable.

Imagine removing a 100-pound chunk of asphalt pavement from the road in front of your home. Placing it on your kitchen table, you and your family would be staring at 95 pounds of small stones held together by five pounds of black glue called asphalt.

Aggregates—stone, sand, and gravel—give the road strength, and asphalt binds them together. Baked to perfection in an oven the size of a school bus, asphalt and stone are blended together in what is called a hot mix.

Ninety-four percent of America's road net is surfaced with asphalt.

In ten years, it is predicted, about 4 million trucks will be hauling double the cargo they do today.

The stones, having been crushed and sorted as to size, are metered into the plant according to a formula developed in the laboratory at a rate of 100 to 600 tons per hour; they are dried, heated, and then mixed with liquid asphalt, which has been refined from crude oil and delivered to the asphalt plant. After the hot rocks and asphalt are blended, the hot mix is hauled by trucks to the construction site of a new road or the repaving of an old one.

Arriving at the paving site, the hot mix is spread over the road and compacted by power rollers—the modern version of an old-time steamroller. It's rolled again and again until it is dense and smooth. Traffic can use the road soon after it has been compacted.

KEEP ON TRUCKING

One of the busiest highways in the world is made with asphalt. Enduring extreme punishment from trucks and cars, the New Jersey Turnpike is a great test case for the durability of any pavement. In 1951, the turnpike became one of the first superhighways in the world to be built from bottom to top with asphalt. From its foundation to its travel surface it was—and remains—an all-asphalt superhighway.

Searching for the ultimate road pavement—one that could be built with the fewest dollars, laid down with the least amount of time, and last the longest—investors backing the New Jersey Turnpike's bonds approved asphalt construction. Fifty years later, having endured brutal loads and high traffic volumes, the original roadway is a living testament to the invincibility of asphalt.

When the New Jersey Turnpike was a new highway, trucking in the United States was in its adolescence. In 1957, just 10 percent of all shipped goods were sent by truck, with the remaining 90 percent sent by railroad. Today, those numbers are nearly inverted, with trucks and asphalt roads carrying the vast majority of freight. In ten years, it is predicted, about 4 million trucks will be hauling double the cargo they do today. As America keeps pace with the world's economy, it will turn to trucks and asphalt.

Deep in the woods in Alabama, a 1.7-mile test track is simulating the Interstate System under unreal circumstances. Trucks loaded to twice their maximum legal weight, tipping the scales at almost 160,000 pounds, chase one another around what looks like a NASCAR track without the grandstands and crowds. American competitiveness in a cutthroat world economy depends on the marriage between truck and asphalt. The Alabama test track is seeing to it that it is a long and happy bond.

THE ASPHALTNET

The Internet carries text messages and digital images. Telephone wires transmit voices. Asphalt roads are the ultimate in communication: They carry human beings and their belongings from one locale to another. A hundred years ago, there wasn't a paved highway anywhere in America. Today, we couldn't live without them.

Asphalt roads are the ultimate in communication: They carry human beings and their belongings from one locale to another.

Everything matters when it comes to the road. Communities prosper or perish depending on their proximity to a highway's location or exit ramp. Our lives depend on the road's condition and maintenance. Asphalt keeps these roads in the best condition and allows for quick repairs.

America's roadways have made us the envy of the world. It is no wonder that the first order of business for a developing nation is to build roads so that commerce can begin to flow. But it's not just about wealth. It's also about strength and security. Our Interstate System was conceived as a way to connect the nation and provide for the smooth flow of troops and matériel for the national defense.

In the supreme example of their worth and flexibility, asphalt roadways can be converted instantly from peacetime uses to tools of security and strength in time of war or terrorism.

War, commerce, national disasters, vocations, vacations, require our hitting the road. The humanitarianism, industrialization, and security of the United States has progressed in lockstep with the development of its paved roads. To truly understand American history is to know the story of America's hot mix-—asphalt.

CHAPTER TWO

Roman Roads

"In the beginning was the road."

THE COLOSSAL LEGEND of Roman roads is exceeded only by the legendary colossal size of the roads themselves. Up to three times the thickness of a modern superhighway and built by the hands of soldiers and slaves, the Roman road was unrivaled for nearly 2000 years.

ALL ROADS LEAD TO ROME

It's hard to say whether the Roman road system grew as the empire grew or the Roman Empire grew as its road network grew. The relationship was symbiotic. Romans revered their roads. Outsiders revered the Romans because of their roads. Using its roads, Rome ruled Europe, northern Africa, and parts of Asia for the better part of 1000 years. Along its 50,000 miles of highways and 200,000 miles of secondary roads, the Roman Empire deployed its armies, quashed rebellions, and sacked and plundered cities. The roads were used for sending their newly taken riches back to the capital, giving additional meaning to the saying "All roads lead to Rome."

Herculaneum was buried some 50 feet deep when Vesuvius erupted in A.D. 79. Its excavation revealed not only intact buildings but the incredibly durable Roman stone roads and curbs.

Rome, with its relatively small percentage of the world's population, was able to exert its influence over a much larger mass of humanity by quickly moving its armies over its roads from one political hot spot to another. The roads allowed Roman army generals to split up their own fighting forces, conquering two different enemies on two different fronts. While in the heat of a battle, a part of a single legion could be deployed over the road network to a developing crisis while the other part of the same legion remained behind to finish putting down the foe there. Flexibility in deployment and mobility over roads made the Roman army lethal. Roman army legions, 5000 soldiers strong—the size of Allied infantry regiments during the Normandy invasion on D-day—were marching 40 miles in a day and 600 miles in a single campaign. Moving in advance of these legions were disciplined road engineers, preparing the roads and bridges for the heavy traffic the military forces put upon them.

Julius Caesar slept in his chariot while it was under way, making the best time possible between conquests. Doing so, he was able to travel over his roads between Rome and the foothills of the Swiss Alps in a remarkable three days. Moving faster than a Roman army or even a dictator in his chariot was the Roman courier. With a well-organized system of post houses, placed about every six miles along the main highways, riders changed weary horses for fresh ones, carrying messages over 100 miles in a day. Napoleon's forces, nearly 2000 years later, were unable to deliver messages any faster. It took the railroads to best the Romans.

Land taxes and donations from Roman citizens were needed to sustain the empire's road building. Contributing money to a road fund in ancient Rome was as noble as donating to the arts and sciences. In return, a road might be named after the wealthy benefactor—an honor of the highest order. If an emperor found a powerful senator had come into sudden riches, he demanded a "gift" to the road fund from the willing or unwilling politician.

Surveyors lined up the road between strategic military posts, key commercial ports, or major areas of population. Working with a crude wooden survey instrument called a *groma*, they laid out stunningly direct routes, rarely deviating except under extreme conditions. Using hills as vantage points and communicating with smoke and hand signals, surveyors made corrections to the course, adjusting their red wooden survey stakes and updating the data to a map. Once plans had been double-checked and charted, the hard labor of building began.

Most of the work fell on the shoulders of the soldiers and slaves. If a citizen of Rome was asked to serve in the army, his burden was heavy—up to 16 years

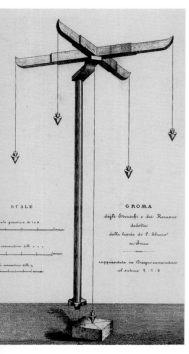

Working with a crude wooden survey instrument called a *groma*, Roman surveyors laid out stunningly direct routes.

for a typical infantryman—but far lighter than that of a slave. Soldiers and slaves began by digging two shallow trenches called zone ditches, marking the outer limits of the road's right-of-way. The zone ditches, as much as 50 feet apart, were for demarcation. The entire 50-foot alignment was cleared of bushes and trees, to protect travelers from ambush by enemy combatants or highway robbers.

Moving in about 10 feet from the outer zone ditches, soldiers and slaves began digging another set of parallel but deeper trenches for the road's primary drainage system. Between these two deep drainage ditches and in the dead center of the survey markers was the travel surface, about 14 feet wide. Preparing the bottom of the roadbed for its heavy stone foundation, workers compacted the earth with ramming rods—wooden poles with weights attached. Sometimes wooden and stone rollers were used to compress the ground, but wood was too light and stone too heavy to make these practical.

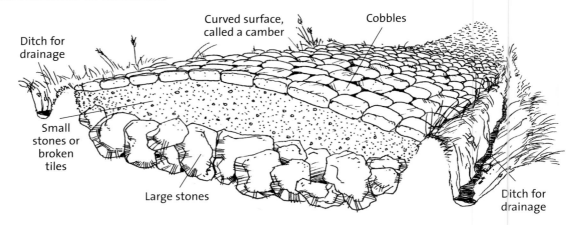

Curved surface,
called a camber

Cobbles

Ditch for
drainage

Small
stones or
broken
tiles

Large stones

Ditch for
drainage

A cutaway drawing of Roman road construction shows the curved surface that forced traffic to the middle for fear of rolling off. The center of the road thus bore the brunt of the wear and tear.

A foundation of flat six-inch-thick stones was then laid on the compressed ground. These rocks were dragged to the work site from a few miles away—sometimes farther if the road was a principal military or commercial road, which mandated higher-quality building materials. With the large rocks in their final position, small stones were wedged between them, preventing them from shifting and roughening the road's eventual travel surface.

Making the second layer, the soldiers and slaves worked with a combination of sand, clay, silt, and broken rocks. This middle layer was compacted as much as possible, and sometimes a lime mortar or slag from a local mining operation was used, forming a solid, water-resistant mass.

The third and final layer was more of the same: sand, smaller stones, tile fragments, and sometimes some additional mortar. The final compacted layer became the road's travel surface in rural areas. If the road was built as a primary thoroughfare in a principal city or was an important military road, more often than not it was paved with a stone surface of some type. Sometimes road builders harvested thin layers of fractured rock, broken by heating boulders with fire and cooling them suddenly, and applied the results to the road as pavement.

TWO THOUSAND YEARS AHEAD OF THEIR TIME

Sitting high above the ground on their man-made earth embankment, roads were first called highways by the Romans. Their highways often had seven-foot-wide shoulders on either side of the main 14-foot-wide travel surface. This foot-high travel area forced everyone to the center of the road for fear of rolling off it and into a ditch. Chariots, carts, and pedestrians straddling the centerline damaged the road's foundation and wore it out prematurely.

Accurately measuring distances between cities on the new roads was vital in gauging travel time for armies and the mail—and for assessing taxes. In order to pay for these monuments of earth and stone, landowners were taxed according to how much Roman road serviced their geographic area. The task of measurement was achieved by a horse and cart rigged with a primitive Greek odometer that may have been the invention of mathematician and inventor Archimedes. Each Roman mile was counted with a single pebble. A full turn of one of the odometer cart's wheels nudged a set of wooden gears. Many nudges later, the wooden gears had been rotated 360 degrees, signaling that a mile had been traveled. This full rotation triggered the release of a pebble from a container into a collection box. At the end of the road, the pebbles in the collection box were counted, each one indicating a mile traveled.

Roman engineers were the first to build road tunnels through rock. It could take a year to dig a few yards by heating the rock with fire and then suddenly cooling it with water, causing the rock face to fracture. Using these arduous methods, they built tunnels as long as 120 feet in length. The Romans also erected equally impressive bridges, sometimes building them of wood so that destroying them during hasty retreats remained an option. Many of the bridges the Romans built of stone are still standing today.

The achievements of the Romans were extraordinary. Using only the power of man and beast, they built useful structures of stone that have lasted for 2000 years. Their determination and strength of will is still to be seen in the aqueducts, amphitheaters, bridges, and roads all over Europe. Many factors made their roads endure: bold design, excellent drainage, a strong foundation, painstaking construction practices, and good-quality stones in the proper sizes and shapes. The Roman roads were the first long-life pavements. Two millennia later, we have Perpetual Pavements—asphalt pavements that resemble the Romans' in that their structures are intended to last indefinitely, with the only maintenance being periodic renewal and recycling of the surface.

The fall of Rome marked the beginning of the Dark Ages in Europe. With no central power to control trade and assure safe passage over the 50,000 miles of Roman highway, the roads plunged into disuse. Communication, trade, and travel along the ancient "interstate" system came to a grinding halt. In Europe, rule by Rome degenerated into rule by many warring fiefdoms, making travel between regions more dangerous. The wooden bridges rotted and fell away, and many of the grand stone roads were taken apart and used by peasants to erect walls between their fields and homes for their families.

It would be more than 1200 years before the art of road building would see its rebirth.

The achievements of the Romans were extraordinary. Using only the power of man and beast, they built useful structures of stone that have lasted for 2000 years.

MOCK-ADAM-IZING — the Colossus of Roads.

CHAPTER THREE

Macadamizing

"But who effected this improvement in your paving?" says Mirabel. "A party of the name of McAdam," is the reply, "but coachmen call him the Colossus of Roads."

Pub. by S.W. Fores. March 1 18
41. Piccadilly London

JOHN LOUDON MCADAM changed the world with his road-building ways, pulling the practice out of its dark ages and ancient traditions, acting as the bridge between the Roman roads and today's asphalt ones. From the end of the Roman Empire to McAdam's day—the late 18th and early 19th centuries—little paving was done in Europe. His concepts were simple and radical—thin, flexible, water-resistant travel surfaces of crushed stone and gravel.

THE COLOSSUS OF ROADS

27

McAdam's determination to rid the world of "wicked ways"—meaning flawed roads, a term he applied to nearly every such surface in Europe—pulled road building out of antiquity. Roads paved with his namesake, macadam, helped to make the Industrial Revolution a reality.

"Few people seem to realize what McAdam did for this country. Had it not been for his roads the industrial revolution could not possibly have taken place for there would have been no means of transport to the new markets that were indispensable to its increased production," said Professor G. M. Trevelyan, author of *British History of the Nineteenth Century.*

In addition to aiding commercial trade, macadamized roads spurred English citizens to get up and out of their hometowns and travel as never before. The stagecoach and horse-breeding industries, grateful for an upsurge in their businesses due to his smooth roads, crowned him the

John McAdam became a household word by revolutionizing the age-old methods of road construction. Here the Scotsman is lampooned in an early-19th-century print.

Before macadamizing, horse-drawn vehicles, people, and animals created unsanitary and uncomfortable travel conditions on London's roads.

Colossus of Roads—a word play on the gigantic ancient statue known as the Colussus of Rhodes. Over 3000 coaches, 150,000 horses, and 30,000 coachmen, guards, and horse keepers were employed as a direct result of the improvements his macadamized roads brought about in the British Isles.

Julius Caesar tried twice to conquer England and Scotland. He failed miserably both times. But when the Romans finally did triumph, they crisscrossed the island with their signature roads. While the island was occupied by the Roman army, livestock, cheeses, fruits, and even oysters flowed over its highways along with light and heavy metals such as tin and lead. Not until McAdam's roads turned transportation around did the British Isles, nearly 1400 years later, meet the levels of trade the Romans enjoyed.

In the days before John McAdam's road-building revolution, commuting over the road was done mostly on a horse's back. Wheeled traffic was too painful to endure. The "wicked ways" were rutted and rugged, causing heavy damage to vehicles. Even Queen Elizabeth complained about the state of her own roads and her royal carriage's bumpy ride, preferring to travel on the back of her royal horse.

For more than 1000 years, kings and queens called for improvements, especially when they were due to travel over a certain section of highway. Royal orders in the form of labor laws required that villagers be put to work repairing and rebuilding roads. But commanding that unskilled subjects go about building roads and making road repairs seldom led to road improvements. They hadn't a clue when it came to making complex repairs—some of the roads being the original Roman ones, whose foundations needed serious reworking.

John McAdam was an unlikely engineering hero. For starters, he was not an engineer. Nor was he a surveyor.

AN UNLIKELY HERO

John McAdam was an unlikely engineering hero. For starters, he was not an engineer. Nor was he a surveyor. McAdam became a trustee of his local turnpike and became so interested in road building that he actually taught himself engineering through reading and observation. Through experimentation, he developed his new ideas for road making and started writing a book. His new methods began to attract attention, leading to his appointment as Surveyor for England's Bristol Turnpike Roads in 1816.

In the same year, McAdam's book, *The Present System of Road Making With Observations Deduced from Practice and Experience*, was first published. It became a bestseller of sorts, inspiring engineers to quote from it in the far corners of the earth.

The Present System of Road Making, which contains no diagrams, is more of a testimonial than a how-to book. It lists postscripts from hearings by the Select Committee appointed by the

House of Commons starting in 1811, including sworn statements from road builders and non–road builders singing McAdam's praises and encouraging lawmakers to require that McAdam's methods be followed. One witness summed up the book as follows.

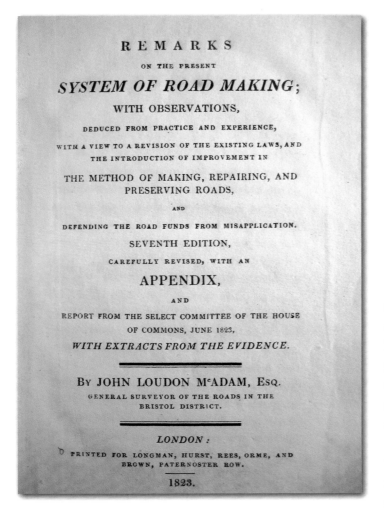

It appears that Mr. McAdam first directed the public's attention to this important fact, that angular fragments of hard materials, sufficiently reduced in size, will coalesce or bind, without other mixture, into a mass of compacted stone nearly impenetrable to water, which being laid almost flat, so as to allow of carriages passing freely upon all parts of the road, will wear evenly throughout, not exhibiting the appearance of ruts or any other inequalities.

McAdam had discovered that both using the right materials and placing the materials correctly were vital to the enterprise of road building. His emphasis on the shape of the stones ("angular fragments") and on stone-on-stone contact made his roads the precursors to some of the premier pavement types of the 21st century.

Looking at the big picture, he advocated that government take a centralized role in road building. As the Romans did before him and the Americans did after him, he understood that it was impossible to build a uniform net of roads if the work were left solely to local jurisdictions.

In short, McAdam believed that the right materials, skillfully applied, would make for a lasting surface; further, that expert administration, maintenance, and repairs would lessen the toll on the beast of burden, the cart it pulled, and the driver who was being charged to travel the turnpikes.

A REVOLUTION IN ROAD BUILDING

The 1700s were the Age of Enlightenment in England and France. They also were the time of a renaissance in road building. Since the fall of the Roman Empire, few real roads had been built, and commerce was difficult as a result. McAdam, along with his fellow Scotsman Thomas Telford and the Frenchman Pierre-Marie-Jérôme Trésaguet, led a revolution in road building that produced vast cultural changes.

In the late 1700s, Trésaguet was busy building the roads Napoleon's armies would travel. Trésaguet is widely credited with establishing the first scientific approach to road building in 1764. Among his innovations was the use of a base layer of large stone covered with a thin layer of smaller stone. Trésaguet's roads made it possible for the Emperor Napoleon to dominate Europe, just as the Romans had done before him.

Thomas Telford's and John McAdam's life spans were nearly identical. The two men were born within a year of each other and died in the 1830s. Both were Scotsmen who were self-taught and passionate about roads. Professionally, however, their practices differed greatly. When it came to road building, Telford was a revisionist and McAdam a revolutionary. Following Trésaguet's philosophy, Telford's roadway designs were more elaborate than the Frenchman's and thicker than McAdam's. More Roman in structure, Telford's required large stones carefully placed into the road's foundation.

Instead of employing the Roman and French methods of placing thick stones in the base of a road and smaller ones at the top, McAdam stressed that only stones of six ounces or less in weight and only those two inches or less in diameter should be placed in his roads. He added that hand-broken six-ounce stones were superior to natural stones of the same weight because they had sharper edges, creating an ideal locking action.

Drainage was one topic on which McAdam and the Romans agreed. If the adage in real estate is location, location, location, the motto for road builders is drainage, drainage, drainage. Emphasizing the need to keep the foundations of his highways dry, McAdam called the travel surface of his roads the roof.

The 1700s were the Age of Enlightenment in England and France. They also were the time of a renaissance in road building.

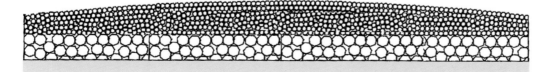

Although Trésaguet and Telford continued to use small rounded stones atop larger ones, McAdam's revolutionary method used two-inch angular stones to create an interlocked structure.

MACADAMIZING AND REMACADAMIZING

On October 8, 1824, *The Times* of London reported the newsworthy event of work crews blocking off Parliament Street in London with temporary fencing, explaining that "Yesterday the workmen began to Macadamize the wide roadway." Eight days, later *The Times* printed, "Macadamizing the whole of the Quadrant of Piccadilly was on Monday enclosed prepatory to taking up the granite pavement lately laid down." Replacing the granite pavements and improving the road with macadam surfaces became so routine that eventually it was a non-event.

McAdam was as meticulous about his road's reconstruction as he was about its construction. "Lifting" a road was the process of repairing a previously macadamized surface. With five men to a crew, the laborers dug down four inches into a road. After they gouged out the old stones and the damaged area, newly broken stones were placed into the road. The process was simple and inexpen-

By the 19th century, London's macadamized streets were smoother, more durable, and easier to clean. The macadamizing of Piccadilly, shown here, was celebrated by *The Times* of London.

sive. Bragging about its practicality, McAdam said the cost of remacadamizing wouldn't exceed a penny a square yard.

After the initial work was completed, the repaired road was reopened to traffic. McAdam prescribed that work details remain behind at the recently lifted area for as long as it took passing carriages to fully compact the newly placed stones. Using heavy brooms, workers continually massaged the loose fragments back into the lifted area, ensuring that the famous macadam surface was formed.

McAdam was as meticulous about his road's reconstruction as he was about its construction.

AMERICANIZING MACADAMIZING

Andrew Jackson seemed an unlikely American hero. He shot and killed another man over a slur directed at his wife. He was known to engage in brawls, leading a British historian to label him "violent, quarrelsome, vigorous, brusque, and uncouth." He was born in a log cabin and, residing in Tennessee, considered himself a Westerner. Detesting the British, he might not have thought highly of John McAdam—as a child of 13 he fought against the British in the Revolutionary War, and his troops later defeated a British army in the Battle of New Orleans in 1812. Then in 1829, General Andrew Jackson became the first president-elect to travel to his inauguration over a federally funded interstate highway, the National Road.

At this point in the nation's history, the National Road was the darling of the country, even though it was considered an "artificial road." A real road was a navigational route across the ocean, along the coast, or up a river. Most artificial roads at the time ran approximately north and south, following the coastline and rivers. There was not a dominant east–west route over the Appalachian Mountains until the National Road. But to Jackson's mind, the nation really began west of the Appalachian Mountains. Opening up the West was his goal, and the new artificial road was proving the best way to do it.

Riding to his inauguration by stagecoach and making celebratory appearances at the inns built each mile along the National Road, the American hero and man of the people was one of the National Road's biggest supporters. The people wanted the road, and he wanted it for them.

During Andrew Jackson's presidency, it seemed that everybody was traveling it—statesmen, celebrities, pioneers, and thieves. Despite the people, cargo, and livestock moving along its way, the most precious cargo on the National Road was the growing nation's infant democracy. The first part of the road that was

President-elect
Andrew Jackson rode to Washington on the National Road, which he knew would open the West. Inns, located each mile along the way, gave him an opportunity to interact with his constituents.

completed went directly through the heart of the nation's population. Facilitating movement and communication between the newly formed states, it allowed the immature nation to grow stronger economically and militarily. So active was the traffic over the road, it was impossible to know whether the nation was pushing the highway west or the fabled road was pulling the country along with it.

As popular and important as the road was, it needed all the help it could get. It quickly began to fall apart. Pioneers and speculators overwhelmed the road. Conestoga freight wagons drawn by oxen hauled supplies to settlers and returned with raw materials from the frontier. Stagecoach companies added to the road's demise by dragging logs behind their coaches when they wanted to slow their descent down the mountain stretches. As drovers herded thousands of cattle, pigs, and chickens to eastern slaughterhouses, the animals ripped apart the road's surface with their hooves.

The root of the problem was in the road's construction, not its traffic. Much of the first section, built between 1808 and 1818 from Cumberland, Maryland, to Wheeling, Virginia, was constructed with heavy stone foundations. The thick and expensive road was impractical for the mountainous route. Not only was it difficult to haul the heavy stones to the road's work sites, but the design meant that once a part of the road's heavy foundation was damaged, it was costly to repair. As a result, most of the damaged sections of road were never mended. With inexperienced hands doing the work and a pathetic maintenance and repair budget of $88 a mile, a more resilient and less expensive surface was needed.

Versed in the ways of macadamizing and under the jurisdiction of the Secretary of War, the U.S. Army Corps of Engineers took control of the second phase of the National Road's construction. West Point–trained engineers replaced the slipshod, unsupervised, and unskilled civilian superintendents who had allowed corners to be cut during the road's first phases of construction. By 1828, with direct military oversight, the civilian contractors were macadamizing a 73-mile section of the National Road between Wheeling and Zanesville, Ohio, giving the National Road the distinction of being the first newly built macadam road highway in the United States.

The National Road's popularity was its downfall. Heavy wagons and thousands of animals quickly destroyed the surface—until macadam was applied.

The macadamizing of the new section of the National Road went so well, it was decided that the heavily traveled and badly damaged section of the road between Cumberland and Wheeling would be repaired and replaced with macadam surfaces. Starting in 1831, Congress again turned to the U.S. Army Corps. Military officers, quoting John McAdam's book *The Present System of Road Making*, communicated to one another by letters about their successes and tribulations along the way to macadamizing the most traveled sections of the National Road.

Travelers witnessed construction crews ripping up the old road. Breaking these larger stones down into two-inch stones, the reconstruction crews were re-entombing the material into the roadbed from which it came. The National Road's old surface was successfully recycled into a macadamized roadway.

Toll-gate houses were erected on
the National Road beginning in 1833,
to collect funds for improvements
such as macadamizing.

Thanks to John McAdam, the National Road began to take on the look of the crown jewel of the Republic.

STILL SOMETHING MISSING

McAdam and his contemporaries who distinguished themselves in road engineering came to some of the same conclusions as the Romans. All emphasized the importance of drainage. All agreed that a strong base was crucial, and that plenty of stone-on-stone contact could make a pavement more durable. These early engineers recognized the importance of the shape, size, and quality of the stones used. They knew that good construction practices were critical.

Still, despite all the advances in road-building theory, materials, and techniques in the 1700s and early 1800s, things were far from perfect. Traffic was light—literally, since most travel was done on foot, on horseback, or in horse-drawn vehicles—but roads did not stay smooth for long. In cities, the accumulation of animal waste on roads was a problem.

Something was missing. That something, as the spiritual descendants of McAdam would learn, was a substance that could bind the stones together, keep the surface smooth and free of dust, quiet the racket of wheels rolling over stones, and make the "roof" of the pavement waterproof. In brief, what was missing was asphalt.

CHAPTER FOUR

Asphalting the Avenues

I stop the noise from the street, so appalling;
Increase the real value on roads,
Prevent dumb brutes from slipping and falling;
While they draw much heavier loads.

38   **ON MARCH 4, 1829,** the newly sworn-in seventh President of the United States, having refused an inaugural parade extravaganza, rode alone on his horse down Capitol Hill. Old Hickory, nicknamed for having the toughness of the hardwood tree, moved slowly along Pennsylvania Avenue and through 16 blocks of cheering women waving symbolic hickory brooms and wearing hickory nut necklaces and men riding horses clad in hickory bark, until he arrived at his new home, the White House. It was a short but messy jaunt.

At the time of President Andrew Jackson's inauguration, America's premier boulevard was neither stately nor grand. It was as full of pigs, chickens, and grazing cows as it was of potential. Constructed in a swampy area, the "soon to be beautiful" avenue sometimes lay under several feet of water. Residents of the time said that "in rainy weather the mules and horses passing up Pennsylvania Avenue would often sink to their backs in the mud."

"I have never seen such a crowd here before. People have come five hundred miles to see General Jackson," wrote Daniel Webster of the

inauguration episode. That day, he made his way through the hordes of well-wishers on a filthy Pennsylvania Avenue; but President Andrew Jackson would be able to claim when he left the White House that he left the street better off than when he arrived. A few years later, in 1832, Pennsylvania Avenue was finally macadamized. It was a start, but the avenue and the nation's capital had a long way to go.

GENERAL GRANT'S GRAND PLAN

Many questioned whether the city deserved the title The Nation's Capital. After the Civil War, federal officials were under pressure to move the reunited nation's seat of power farther west—to Ohio, Indiana, Illinois, Missouri, or Kansas, anywhere but where it was.

The problem was that Washington, D.C., wasn't instantly prestigious or commercially successful, as had been anticipated. At the center of the dilemma were the city's streets. More visions of the future than actual avenues, the planned grand streets were merely designated by survey markers at that time. The avenues that did exist dead-ended into pastures, earning the town the ghostly title of "city of streets without houses."

In 1840, a grand but rutted Pennsylvania Avenue was not yet the beautiful thoroughfare that had been promised.

Following the Civil War, the streets of Washington, D.C., were in poor condition. Soldiers returning from the war-torn South told residents that their city's streets were as rough as the ones they had encountered while fighting in Virginia, where it took six horses and an entire day to pull a howitzer cannon just a few miles.

In 1870, President Ulysses Grant helped to save the nation's capital by launching a plan for paving its streets. Approving a scheme to find the best surface for the city's streets and successfully reverse Washington's reputation as a backwater capital, President Grant placed his hopes in the hands of a friend, Alexander "Boss" Shepherd. Attacking Tiber Creek and laying a pipe in it, Boss Shepherd's gangs converted it from a latrine to a modern sewer. By backfilling the canal with dirt and paving a road over it, the project made way for one of the city's main boulevards, Constitution Avenue. But that was only the beginning of an unprecedented $6.25-million plan. The end goal called for 74 miles of sewers, gas and water lines, and 50,000 trees. At the heart of the agenda was the paving of nearly 200 miles of city streets.

As troops paraded to celebrate the end of the Civil War, Pennsylvania Avenue was one of the few decent streets in the nation's capital.

On a roll and searching for the best pavements a city could buy, Boss Shepherd and his entourage took a road trip in the spring of 1872. Meeting with officials of Philadelphia, New York, Brooklyn, Boston, and Buffalo, they determined which pavements they would experiment with. Upon their return, it was decided that the entire city of Washington, D.C., would be a sort of test case of travel surfaces to be constructed and then observed for durability and practicality. Their plan called for paving 93 miles with cobblestone and stone macadam surfaces, 58 miles with wood block, and 28 miles with asphalt.

By 1874, Boss Shepherd's efforts were taking hold. Like the Civil War itself, the project was filled with tribulations and triumphs, but within three years the ambitious scheme "lifted Washington out of the mud." The pavements were credited with the success, but which pavement type would last the longest, cost the least, and do the most for the city was still an unknown.

Wooden planks formed the "pavement" for San Francisco's Sacramento Street in 1854.

THE SEARCH IS ON

Washington, D.C., was in bad shape but in good company. Its sister cities in Europe—London, Paris, and Berlin—were also searching for the ideal road-paving materials. In a futile effort to make their living environment cleaner, healthier, and more hospitable, many government officials, civil engineers, and entrepreneurs worldwide were exhausting a long list of materials to find the perfect material for paving roads. The brightest men from the wealthiest nations were always falling short of their goal of a resilient, long-lasting travel surface that could be easily maintained.

The brightest men from the wealthiest nations were always falling short of their goal of a resilient, long-lasting travel surface that could be easily maintained.

The rewards for finding the cure were obvious. An 1882 description of San Francisco states, "Market Street is swept every day, but every evening from two to seven o'clock, when it is most used by the pedestrian class, it is the dirtiest thoroughfare in the city. The sweepings, ground into fine particles, saturated with all the unclean and unhealthy things, sometimes including the glanderous droppings of diseased horses, as infectious as the small-pox or leprosy, are scattered by the west wind over all the pavements, inhaled in the nostrils, eyes, throats, of everybody on the street, not merely to their great discomfort, but to the constant peril of their health."

Besides eliminating the plague of dust and dirt, the right pavement would decrease the cost of consumer goods by reducing hauling charges and wear and tear on horse and cart, increase the number of products and types of produce available, make the city safer from fire by shortening firefighters' response time, increase "social intercourse," and improve public health by making it easier to clean up droppings left by the beasts of burden.

In the latter half of the 1800s, Berlin's population and traffic counts were doubling every 20 years. Struggling to cope with their growth, Berlin Public Roadway authorities were experimenting with pavements. "With extreme strictness," they were testing the stone blocks of Belgium, wooden blocks from Sweden, and asphalts from Switzerland, Italy, and Germany. Citizens of Berlin, London, and Paris as well as Washington were wrestling with the same issues and discovering the same shortcomings. Stone and wood pavements weren't working. Asphalt, on the other hand, looked promising.

STREETS OF STEEL

In a try-everything approach to the search for a durable road, Cortlandt Street in New York City, not far from Wall Street, was paved with iron in the mid-1800s. Harder than stone, the road was durable, promising not to rot like wood. Otherwise, it was a terrible surface. Slippery when dry, in the rain it became deadly for horses pulling heavily loaded wagons. Sharp hexagonal rises, about an inch in diameter, covered every bit of the road surface. The intent was to give the animals a firm footing. Instead, the hexagons tugged and twisted at the cleats on horseshoes. When a horse fell, it tore its knees on the jagged surface.

Rough and loud, Cortlandt Street's metal lane was replaced with a slightly more humane stone surface shortly after it was laid down.

STONES AND BLOCKS

Almost as hard as steel and much gentler to a horse was stone. Bigger than a nugget and smaller than a boulder, cobblestones were the mainstay pavement for cities in Europe for hundreds of years. Readily available, they were cheap; but, random in size and shape, they wore smooth too quickly. The travel surface of a cobblestone was to be no less than four inches and no more than eight inches in diameter, and it was to be set between five and ten inches into the ground. As cobblestones were used more widely, they became scarcer, and the specifications became less strict. After a time, in an attempt to manage quality, some cities outlawed their use on the main boulevards, relegating them to freight yards, alleys, and gutters.

Belgium blocks of stone picked up where cobblestones left off.

Cobblestones had been used for centuries, but their shortcomings led to trials of other components.

Belgium blocks were more uniform in size than cobblestones, but passing wagons, carriages, and horses' hooves created an unbearable racket.

First used in Brussels, Belgium, they were becoming the pavement of choice by the early 1800s, when European and American cities were laying them down with reckless abandon. Looking like clunky truncated pyramids before they were placed in the road, Belgium blocks usually had a surface of five or six inches long and were typically seven to eight inches deep. Used as early as 1849 on Pennsylvania Avenue, and shortly after that in New York City, Belgium blocks were praised for keeping dust and mud to a minimum.

Making for a rough but durable surface, the tough blocks lasted more than 15 years, even with heavy wear. Squeezing life out of their investment, some cites restored their Belgium blocks by pulling them out of the roadway, grinding them down, and placing them back into the street, proclaiming them as good as new. Though it was expensive, disruptive, and time-consuming, this early form of road recycling had a major advantage: it doubled the life of the street's surface.

The busiest boulevards in Paris in the late 1800s had traffic counts as high as 42,000 horses and carriages a day. Belgium blocks in these parts of the city were blamed for insufferable noise. Residents along these grand avenues hated the racket. Riders and passengers suffered as their internal organs were shaken. Heavy traffic caused Belgium blocks to shift about, making the surface uneven and rutted. One observer of the times noted that they made for a "greasy and slippery surface when damp." When a person's prized possession, the horse, took a tumble on wet blocks, the animal was often seriously injured, or worse.

CLAY BRICKS

Quarrying granite for the making of Belgium blocks was expensive. By contrast, manufacturing clay bricks was laborious. The wrong clay or a misjudgment in the burning of the bricks during manufacturing reduced what looked like a solid brick into red dust with the first passing carriage wheel.

Upon harvesting from riverbanks and fields, the right clay or "stiff mud" was hauled to a brickyard, where the raw material was passed through rollers, flattened of lumps, and dumped into containers where augers mixed it with water. After a thorough mixing, the clay was fed into a forming machine capable of producing an endless clay band two inches high and four inches wide. This elongated clay bar was finally cut up into eight-and-a-half-inch-long bricks.

The moist bricks were trucked to a drying house, where they sat for days. Eventually they were moved to the ovens, where a tricky process of burning and drying the bricks began. The heat in the brick kilns was slowly raised for a week

Bricks baked from clay were easy to lay down (left) and could be set in many patterns (lower left). But if not made to strictest standards, they could crumble or chip.

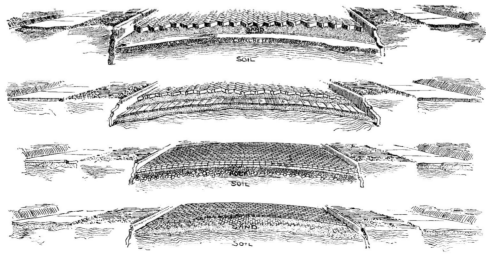

or longer. Heating too fast caused the bricks to become brittle, lose their shape, and turn into worthless piles of dust. Once the last of the dampness was sucked out of the bricks—evident in that no more steam could be seen escaping their surfaces—the burning began. The ovens were brought up to a hellish 1500 to 2300 degrees Fahrenheit. The more impurities in the clay, the higher the temperature needed. The ovens were kept at this level for a week to 10 days until the burning was complete.

Standardization in building materials was becoming common practice by the mid-1800s. Most clay bricks were eight and a half inches long, two and a half inches thick, four inches wide, and weighed five to seven pounds. A contractor might need 3 million bricks for a ten-mile-long, two-lane-wide highway.

Bricks reduced dust and mud and could be easily laid down, cleaned, and repaired. But they were slippery when wet and their quality was inconsistent. Some bricks in the road might be solid, while others crumbled, chipped, and wore down under moderate levels of traffic. Once one weak brick broke, the stronger bricks around it were crushed under passing carriage and wagon wheels.

Popular in the Netherlands in the early 1800s, clay bricks made one of their first appearances as a road building material in the United States in Charleston, West Virginia, in 1870, and became popular in the Ohio Valley area. If a city was lucky, a clay brick street might last 10 years.

WOODEN BLOCKS

Wooden blocks were quieter than brick but tended to decompose, even when coated with coal tar or oil.

Slightly longer than a clay brick and lighter than a Belgium block was the wooden block. These were used in Russia centuries before arriving in New York in the 1830s. Beyond logic, they were used repeatedly for the next half century despite their epic failures.

Wherever they were used, decomposition was an issue. Dry rot in the desert and normal decay in moist climates turned wooden blocks into a fungus-ridden and putrid pavement. Determined city planners kept tempting fate by coating them with coal tar or oil. Health officials worried. But, aside from their tendency to decay, wooden blocks had their good points.

More popular in the United States

Wood-block roads
became uneven when
the blocks rotted and
were pummeled by
traffic, as seen in
these drawings of
worn surfaces.

than in Europe, they were used heavily after the American Civil War. At first they were produced as round "blocks" of cedar. Capturing dirt and moisture where their rounded edges left openings, they quickly rotted. In 1848, an attempt to remedy their rotten ways brought on a patented rectangular block, called the Nicholson. Unremarkable in appearance, it was a square version of the round block and suffered the same fate as its round brethren. Quickly succumbing to wear and tear, the avenues paved with wooden blocks were torn up and repaved often. In Paris, a road builder was obligated to return to the job site every six years and lay a new wooden road.

Each type of wooden block had its own set of nuances, so city officials and road builders had to acquire the expertise of master carpenters. Understanding the nature of cypress, cedar, hemlock, oak, white pine, cottonwood, mesquite, redwood, and others was the only way to understand how a block would fare in the street. Once a city engineer selected the type of wood, a block press in the mills converted the raw lumber into a pile of 240,000 wood bricks each day. Similar in shape to a clay brick, a typical wooden block was about three inches thick, five inches wide and nine inches long. Unforgiving, the blocks could vary no more than a sixteenth of an inch from original specifications. Their high rejection rate and low durability made wooden blocks an expensive endeavor for taxpayers.

A wooden paving brick's best hope for longevity was to be treated with creosote, coal tar, or asphalt. The wooden blocks, piled high along the side of the street before placement, were drenched with as much as 22 pounds of oil for each square foot of wooden pavement laid. Once the blocks had received a thorough and expensive coating, workers carefully placed them to prevent scratching, denting, or chipping. The slightest damage to a wooden block rendered it useless. One side benefit to the oils and coal tar—besides extending the life of the wooden road—was that it muffled the noises from horseshoes and hard carriage wheels.

Leon Malo, who laid the first successful coat of asphalt on a European

Their high rejection rate and low durability made wooden blocks an expensive endeavor for taxpayers.

avenue, declared that in Paris, "wood is an embarrassment for public hygiene, asphalt is an auxiliary thereto." About the road tests in Germany, he said, "the vast experiment made at Berlin, with compressed asphalt has succeeded beyond the expectations of its originators and its users," adding that "In London, the success of asphalt may be considered as almost complete." Around the world, city planners were beginning to see that the future of road pavement belonged to asphalt.

ACCIDENTALLY ASPHALTING MACADAM

Rocks filled with asphalt were first discovered and gathered for commercial gain in Europe in 1720 in Neuchatel, Switzerland. Such rocks were later uncovered in France, Germany, and Italy. Europeans at first used the rocks for asphalting their flesh wounds, roofs, and boats. Asphalting their roads wouldn't come for over 100 years.

By 1870, Londoners had paved Threadneedle and Cheapside streets with rock asphalt.

As early as 1802, rock asphalt was being used in Europe, but as a building material its uses were limited to the paving of sidewalks, bridge decks, and the floors of some commercial buildings. At the time, applying asphalt to small surfaces was time-consuming, and paving the vast surface of a road made no sense until it happened accidentally.

In Switzerland, chunks of rock asphalt, piled high onto carts for transportation, fell from the overloaded carts onto the surface of macadam roads they were passing over. Acting as rollers, the wheels of the wagons compacted the asphalt into the stone road. The combination of compaction, asphalt-impregnated limestone, and summer heat created an impressively—albeit accidentally—asphalted macadam pavement. Taking the asphalt-laden stones and breaking them up as they would for any other macadam road, Europeans intentionally began combining asphalt with John Loudon McAdam's construction principles.

Major cities had to address the problems of heavy traffic by the 1880s, as seen here in London.

ROCK ASPHALT

European city officials were slow to take to rock asphalt's hard and shiny black surface; stone and wood still looked like the materials of choice. Not until Leon Malo, a French engineer, successfully laid down rock asphalt on the rue Bergère in Paris in 1854 did the Continent take notice. By 1870, Londoners had paved Threadneedle and Cheapside streets with rock asphalt. Berlin, not to be left behind, began paving streets with asphalt too. Europe started to tear up its

cobblestones, Belgium blocks, and wooden pavements and give their boulevards a black top. By 1888, Paris had 20 miles of asphalt avenues and London was not far behind, with 15 miles of the new type of pavement.

Though quite durable, rock asphalt's biggest drawback was its slipperiness, especially when wet. Limestone in the asphalt gave the rock asphalt a slickness that worsened in a moist atmosphere, which reduced the friction between the road's surface and the metal shoes of the passing horses. As a result, Belgium blocks, bricks, or even wood were still used on the hilly parts of city streets to give the steeds and wheels a better grip.

In the United States, over 5000 miles away from Europe's rock asphalt deposits, it was expensive to import both the asphalt and the rocks that contained the asphalt. Only the coastal cities could justify importing and laying rock asphalt as a pavement, because hauling it overland after a long voyage at sea was cost-prohibitive. Boston, New York, and New Orleans used the material, but only in limited amounts.

Nearly eliminating dust and mud, much quieter than Belgium blocks, and easier to clean of horse manure than any other surface, the rock asphalt pavements became fashionably upscale. So expensive was the desirable asphalt that only the grandest boulevards and avenues of the wealthiest citizens were paved with it.

The Champs-Élysées was just one of the streets of Paris that had been asphalted by the 1880s.

"ARTIFICIAL" ASPHALT'S FAILURE

In 1838, the City of Brotherly Love was experimenting with rock asphalt. Protected from rains that would make it slippery, the sidewalk of the front portico of the Philadelphia Merchants Exchange was paved. But not for another 30 years was asphalt used to pave a road in the United States.

Copying their Parisian counterparts, New York City experimented with an asphalt-like pavement on its most prestigious thoroughfare, Fifth Avenue. Unfortunately for the affluent residents of the avenue, this unproven "artificial asphalt" was a bust. On Saturday, October 9, 1869, *The New York Times* ran a front-page story about the disaster of the black dust. A sympathetic journalist reported that "the residents of Fifth Avenue have suffered immeasurably. . . . Many of the Fifth Avenue residents declare that they are willing to pay an additional assessment to cover the cost of removing the asphaltum, which, it was promised, should make their street the model thoroughfare in the City."

The writer continued that "small heaps of dust are found on the stoops and window sills, which await only a breath of air to transfer them from house to house . . . the supply seems to be never ending. Inside the houses mantels, curtains, carpets and furniture generally are continually receiving coatings of dust which are fast working the destruction of articles . . . and the countless fancy

Part of Fifth Avenue in New York was paved with an early version of asphalt in 1869, but it made a disastrous mess and was soon removed.

ROEGE N.Y.

adornments that are to be found in such profusion in these abodes of the wealthy." After three weeks, the road had already worn through to the Belgium blocks it was laid over. Endorsing natural asphalts, the reporter condemned artificial asphalts, saying they should "be rigorously excluded."

It would be decades before "artificial" asphalt would be accepted.

A CHEMIST'S BIG SUCCESS

"In the history of road building in modern times, two names stand out like the sun among the stars—John Loudon McAdam and Edward J. DeSmedt. The former holds the same relation to the improved rural highway that the latter holds to the improved modern city pavement," declared Amzi Lorenzo Barber, a man who in the late 1800s built a natural-asphalt empire called the Barber Asphalt Paving Company. Across the Hudson River in Newark, New Jersey, one year after the calamity on Fifth Avenue, DeSmedt was responsible for the first successful commercial application of asphalt on a city street in the United States.

City Hall in Newark, New Jersey, was the site in the 1870s of the first street in the U.S. to be successfully paved with asphalt.

Edward J. DeSmedt was a Belgian-born chemist who, after attending the chemical school of Brussels, moved to the United States in 1856. He had studied the successful uses of natural rock asphalt in Europe and the famous road pavements in Paris. Before striking the right chord with his mixes, DeSmedt had experimented with the waste of coal plants, called culm, in an attempt to reduce the cost of rock asphalt. He also tried ridding rock asphalt of its slippery reputation by adding sand as an abrasive. He, like others, had been unsuccessful in working with minerals in an attempt to make wooden blocks last longer.

In 1870, however, he created what was needed—a natural asphalt for the road, made with the best materials. On the heels of New York City's failure and with a show of confidence, DeSmedt laid down an impressive sample of his newly concocted lake asphalt in front of his client's place of work: Newark's City Hall. It was a small job but a big step forward for asphalt paving in the United States.

DeSmedt's recipe for success was different from the

typical natural asphalt of the day. The natural asphalt he used was taken from a lake, not a rock. Most significantly, his was the first hot-mix asphalt known to be used on a surface of a road. In Europe, the asphalts were cold mixes: rocks taken from the earth, broken apart, and pulverized onto the surface of the road, the same way a typical macadam road was built. The only heat in the cold-mix process was from the sun, which helped meld the asphalt and stone into a continuous surface. In Newark, DeSmedt intentionally heated natural lake asphalt and then, after churning a sand-powdered limestone into it, he softened his hot-mix asphalt with a residue of petroleum. It was a natural asphalt but an imitation of the European rock asphalt that had already proved itself.

The success in Newark gave DeSmedt a reputation as the man to employ if seriously thinking about paving a road with asphalt. After receiving patents for his mixes and laying down more of the material in Philadelphia between 1870 and 1874, DeSmedt set his eyes on bigger contracts. He was just the innovator the planners in Washington, D.C., were hoping would bid on a highly visible and innovative large-scale asphalting job—the paving of Pennsylvania Avenue between the Capitol and the White House. The risks were high, but so too were the potential rewards.

TRINIDAD LAKE ASPHALT

DeSmedt got his lake asphalt from the Caribbean island of Trinidad, 2000 miles from Newark, New Jersey. Petroleum rising to the surface of Pitch Lake, a 115-acre volcanic lake, sat in the open air. In one of Mother Nature's personal refineries, the lake of exposed petroleum had lost its lighter portions to evaporation by the hot tropical sun. What remains after gas and kerosene are refined in a man-made refinery is asphalt.

The asphalt from Pitch Lake was dubbed with an appropriate name: Trinidad lake asphalt. The lake's asphalt is dense enough to support the weight of people, horses, and even heavy equipment, but it is viscous enough so that a heavy object left standing on its surface will eventually disappear into the asphalt lagoon. This is a typical property of liquid asphalt, and it is referred to as viscoelasticity.

Workers harvesting the asphalt walked freely across the top of the lake's warm surface. Digging with their pickaxes and shovels, they filled wicker baskets, horse-drawn carts, and eventually railcars with Trinidad lake asphalt. Returning to the exact location just two days later, workers were able to pick up where they left off,

The island of Trinidad, off the coast of Venezuela, was DeSmedt's source for lake asphalt.

knowing the "bottomless" lake had already replenished the excavation site.

    After the lake asphalt made its way from the lake to waiting sailing ships, it was dumped into the hulls of the ocean-going vessels for its long trip to either the United States or Europe. Unfortunately, during the ship's trip, the asphalt in the cargo hold would conform to its natural state—a monolithic mass—as if it were back in the volcanic lake bed. When the ship arrived in port, dockhands had to use the same tools to get the asphalt out of the hull of the ship as their counterparts back in Trinidad had used to excavate the material from the dead volcano: pickaxes and shovels.

Trinidad lake asphalt was pickaxed from the source, and loaded into wagons and railcars for transport to ships sailing to the U.S.

Rock asphalt from Europe was slightly more expensive in the U.S. than lake asphalt because of the costs of shipping it an additional 3000 miles from inland Europe. The European rock asphalt was laden with heavy stones, adding to the shipping cost. The fact that Trinidad lake asphalt was relatively pure, with the exception of a few sticks and stones per ton, combined with the fact that horses and carriages were not as prone to slipping and sliding on city streets paved with it, helped make Trinidad the source for nearly all of the world's asphalt between 1875 and 1900.

ASPHALTING WASHINGTON'S PLACE IN HISTORY

In 1876, Washington, D.C., was cleaning up its act for the nation's 100th birthday. Pennsylvania Avenue, the city's main parade route, had become an embarrassment. Five years earlier, undaunted by the facts about rot and disease, city officials had ripped up the avenue's Belgium blocks, replacing the entire stretch between the Capitol and the White House with a hardwood floor. Predictably, Pennsylvania Avenue was awash in rot, scum, and splinters in less than four years.

Encouraged by DeSmedt's success in Newark, Congress, with President Grant's urging, agreed to mop up the slop and lay down a new and promising asphalt surface. Passing an act to replace the rotten wooden pavements with the "best known" materials, Congress created a special commission of three to oversee the work. After receiving over 40 proposals for Belgium blocks, wooden blocks, coal tar, and various asphalts, they awarded only two contracts.

Pennsylvania Avenue became America's test road for pavements.

Pennsylvania Avenue became America's test road for pavements. It had been a dirt road for decades before it was macadamized in 1832. It was paved with cobblestones in 1849 and wooden blocks in 1871. Finally, with the military's oversight, it was going to be given its first coat of asphalt. From the steps of the Capitol building to Sixth Street, the avenue was paved with Neuchatel rock asphalt. From Sixth Street to the White House, DeSmedt's patented Trinidad lake asphalt paving was laid. It was a public beauty contest. The civic judging came down in favor of the lake asphalt because of the rock asphalt's slickness. In a further step, the rock asphalt placed on one portion was replaced with Trinidad lake asphalt, and the substance was written into the road paving specifications in Washington. The city's engineers also developed and standardized tests to measure and control the properties of asphalt.

Pennsylvania Avenue, paved with Trinidad lake asphalt, was said to have "a

surface as smooth as a billiard table and as clean as chiseled stone." One local writer boasted that "Washington is today universally known as the best paved city in America, if not the world."

In the 1880s, Washington, D.C., led the world in asphalt paving, with over 70 miles of avenues and streets paved with the now famous lake asphalt. Having gone from a national embarrassment to a world model, the city became an international destination. Even experimental white asphalt was laid down. Locals teased tourists that it was possible in the heat of July and August to cook on the white avenues—a story reported as gospel by the press.

Talking about his days as a youth in Washington, H. L. Mencken recalled the most exciting attraction in the city: "The greatest of them, in that era, was not the Capitol at the end of Pennsylvania Avenue, nor even the Washington monument but asphalt streets. Asphalt was then a novelty in the United States, and Washington was the only city that could show any considerable spread of it."

Washington was finally staking its claim as a great capital city along with its rivals around the world. Property values were rising sharply, and locals, tourists, and dignitaries were raving about the city's asphalt. "Since the surface of these asphalt pavements . . . the noises of hoofs and wheels are almost entirely muffled, while a pleasant and agreeable footing is offered to horses," marveled a local. Significantly, there was no more talk about relocating the nation's capital.

CHAPTER FIVE

The Asphalt Tycoon

"Having given thirty years to the paving of city streets—during which time the Author has been identified with the laying of more pavements, or the furnishing of material therefor, than any other person in the history of the world, he ventures to submit his views on the building of country roads for the considerate judgment of all who are now trying to solve this important problem in engineering."

—Amzi Lorenzo Barber

WHEN AMZI LORENZO BARBER graduated from Ohio's Oberlin College in 1867, he contemplated following his father into the ministry. Instead, he took a professorship of philosophy and a seat on the board of trustees at the newly formed Howard University in Washington, D.C. Founded for newly freed slaves and blacks who had been born free, Howard was one of the first true opportunities for African Americans to get a college education.

In 1873, after leaving Howard University, Barber purchased 40 acres of the university's grounds and, along with his brother-in-law, created a subdivision of 41 homes. The land was later turned over to the city of Washington and became a center of African American intellectual and cultural life.

Barber's foray into real estate and his experience as an entrepreneur led to his visionary insight into the potential of asphalt and the improvements it could generate in American cities. At the turn of the 20th century, *The New York Times* reported, "It was the inquiry conducted by the United

Smooth and durable lake asphalt on Fifth Street in Washington is an example of the many miles of pavement applied by Amzi Barber's company in the nation's capital.

States Government into the best methods of paving the streets of the capital which turned his attention to asphalt. Pennsylvania Avenue was to be made into the finest thoroughfare it is today and the authorities decided the best material was asphalt, which in 1876 was an entire novelty. Mr. Barber watched its success, and saw what immense possibilities lay in its development, and two years later forsook the real estate business for the manufacture of asphalt pavement."

Barber formed a business called A. L. Barber & Company at the age of 35 and incorporated it five years later, with the firm's legal name becoming the Barber Asphalt Paving Company. By this time, Barber was already busy covering many miles of streets in the nation's capital with asphalt.

The more experienced the men of the Barber Asphalt Paving Company became at working with asphalt, the more they separated themselves from competitors. It was common for a public road-building job to draw a half-dozen bidders. The inexperience of most competitors in this fledgling field was one of the Barber company's greatest contracting advantages. But that wasn't the only advantage Barber enjoyed—the city officials responsible for hiring and overseeing the asphalt paving often depended on Barber's firm to guide them through the steps. The safest bet at the time was to use Trinidad lake asphalt, supplied and applied by Barber's company.

The safest bet at the time was to use Trinidad lake asphalt, supplied and applied by Barber's company.

When the competition finally did improve their techniques, Barber's company challenged them by guaranteeing its work. During the late 1800s, Barber set the cost of paving at about three dollars a square yard. Even if competitors could match this low price, Barber further challenged them with a ten-year warranty. The Barber Asphalt Company's guarantee made it easy for a municipal engineer to choose the company to apply asphalt to his city's avenues with little or no fear of losing his job for having made a bad decision over contracting. If competing contractors lowered their prices to meet Barber's, it was a given that they would have to cut their guarantee in half, to five years. If they matched the guarantee, they would have to raise their prices. Barber Asphalt had its rivals over a barrel.

Asphalt laid down by the Barber Asphalt Paving Company in 1882 showed little wear when cut out six years later.

On the supply side of the business, Barber's Trinidad Asphalt Company was in near-complete control of the source. It consistently delivered asphalt for less than half the cost of its competitors—and sometimes for far less than half. So popular and successful was Trinidad lake asphalt, it became the standard in the industry. Fire chiefs claimed their horses did not slide on it and city engineers

gave testimonials to its durability. Municipalities began specifying its use when drawing up contracts for public work.

It was obvious that asphalt was more than just a promising material. Around the country, demands for the substance and for companies to apply it brought more and more contractors into the asphalt business. Unwilling to see his dominance diminished, Barber moved to block out competitors however he could. Calling in political power wasn't unusual for him, as Senators, Congressmen, and even members of the Cabinet were on his call list.

In the late 1880s, the competing North American Asphalt Company was cutting into Barber's near monopoly by going west. In Utah, they were developing a promising 640-acre deposit of rock asphalt. Making the find even more threatening to Barber was the fact that it lay near the Denver and Rio Grande Western railroad—so hauling the asphalt to the lucrative East Coast markets would be that much easier. Encouraged, the firm began looking for additional sources of rock asphalt nearby.

The discovery of rock asphalt was promising, but there was a catch: It sat on

On the cliffs, along the beaches, and even in the water of the Santa Barbara coastline in California were some natural asphalt deposits.

an Indian reservation. Undaunted, the North American Asphalt Company moved to secure the property by negotiating a lease of some 5000 acres. A turf battle erupted between the 988 Indians who wanted to lease their lands to the North American Asphalt Company, a U.S. Congressman motivated to see a transaction made, and the President's Secretary for the Department of Interior who was refusing them the opportunity.

Supporting the interests of Barber, the Secretary of the Interior ruthlessly turned on the Indian tribe, proclaiming that the lands the Indians were calling their reservation had never been ceded to them as a permanent home and that they had no right to the property. Proving the power of the pen and a bureaucrat, he warned them that they didn't need all of their two million acres and that perhaps just 200,000 acres would be enough. If not for an administrative change due to Presidential politics and the fact that discoveries of asphalt in other parts of the country and the world were coming to light, the battle would have raged on.

A postcard from 1910 extolled "pure natural lake asphalt." Note the steamroller in the right rear.

The other natural asphalt deposits all had limitations that kept Barber and his company in control. On the cliffs, along the beaches, and even in the water of the Santa Barbara coastline in California were some natural asphalt deposits. They were known to locals and seemed promising for the national market, but the asphalt was an expensive 3000-mile train trip or an even more arduous three-plus-month voyage by ship from the lucrative East Coast markets. In Kentucky, asphalt was discovered in remote and hilly areas. Since these were hard to reach on horseback, not to mention with wagons, rail lines needed to be built before these deposits could be mined.

Regardless of his near monopoly in the asphalt industry, Barber continued to throw himself into his work, traveling an average of 1000 miles a week by rail and spending about 120 nights a year on sleeper trains. He claimed to have covered 80,000 miles of ocean and 400,000 miles over land during the first four years after incorporating his company. In successful pursuit of profit, he traveled in style, logging most of his miles in plush Pullman train cars. When plowing the seas, he often used his private steam yacht, anchored near his homes on Staten Island and Fifth Avenue or his corporate headquarters in New York City.

At the close of the 19th century, Barber Asphalt had placed approximately 24 million yards of Trinidad asphalt on nearly 1500 miles of roadway. Nationally, over $50 million worth of asphalt pavement had been laid over streets in 80 cities around the country. Nearly every yard of it had been supplied, and more than

At the close of the 19th century, Barber Asphalt had placed approximately 24 million yards of Trinidad asphalt on nearly 1500 miles of roadway.

half of it placed, by the Barber Asphalt Paving Company. It took some 30 other companies, almost always applying his Trinidad lake asphalt, to lay the other half. Barber was the leading stockholder, director, and officer of both the Barber Asphalt Company and the Trinidad Asphalt Company, putting his net worth at an estimated $7 million by 1895. Headquartered in New York, the firm had offices all around the United States.

"The Men Who Manage the Barber Asphalt Company" was the title of an engineering magazine's story in 1896 that claimed that Barber's company was "one of the greatest business organizations in America." On its face, the statement was true.

STEAMING AHEAD

Asphalt was very good to Amzi Lorenzo Barber. As a result of his success in brokering, supplying, and paving with Trinidad lake asphalt, he lived a life of wealth and fame. Far from following in the footsteps of his preacher father, he moved freely from his mansions to his yachts and to yacht clubs near home and abroad. He supported the arts, and was listed in the Social Register.

Barber bought not only mansions and yachts but the entire company that made this fancy auto.

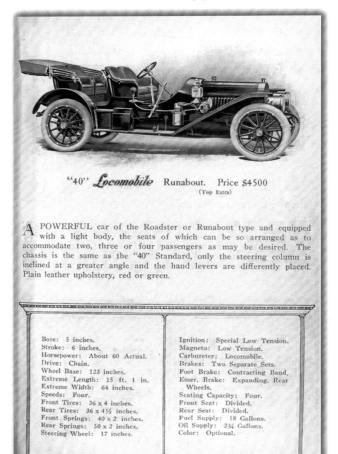

"40" *Locomobile* Runabout. Price $4500
(Top Extra)

A POWERFUL car of the Roadster or Runabout type and equipped with a light body, the seats of which can be so arranged as to accommodate two, three or four passengers as may be desired. The chassis is the same as the "40" Standard, only the steering column is inclined at a greater angle and the hand levers are differently placed. Plain leather upholstery, red or green.

Bore: 5 inches.	Ignition: Special Low Tension.
Stroke: 6 inches.	Magneto: Low Tension.
Horsepower: About 60 Actual.	Carbureter: Locomobile.
Drive: Chain.	Brakes: Two Separate Sets.
Wheel Base: 123 inches.	Foot Brake: Contracting Band.
Extreme Length: 15 ft. 1 in.	Emer. Brake: Expanding, Rear
Extreme Width: 64 inches.	Wheels.
Speeds: Four.	Seating Capacity: Four.
Front Tires: 36 x 4 inches.	Front Seat: Divided.
Rear Tires: 36 x 4½ inches.	Rear Seat: Divided.
Front Springs: 40 x 2 inches.	Fuel Supply: 18 Gallons.
Rear Springs: 50 x 2 inches.	Oil Supply: 2¾ Gallons.
Steering Wheel: 17 inches.	Color: Optional.

Yachting was a passion for Barber. He and his family cruised comfortably along the eastern seaboard and went as far abroad as ports of the Mediterranean. Having a penchant for being at the vanguard of industry, Barber purchased one of the first steam-powered yachts in 1902. It was just the beginning of his fascination with steam engines.

So convinced was Barber that steam engines were the future, he and a partner bought "factory, patent and parts" from the twins who later built the Stanley Steamer. Barber and his partner founded what became known as the Locomobile Company of America in 1898. The coinage *locomobile*, Barber's idea, was intended to imply that the vehicle was part locomotive and part automobile.

Ironically, the steam cars were only successful on smooth city streets, and the Locomobile Company soon joined those manufacturing the more popular gasoline engines.

CREATING THE ASPHALT TRUST

Asphalt seemed to run through Amzi Lorenzo Barber's veins. Despite his love of yachting and his endeavors with automobiles, it was asphalt that drove the man and was the source behind his successes. With a take-no-prisoners approach to winning business and increasing Barber Asphalt Paving Company's market share, Barber became the Asphalt Tycoon.

At first, his company brokered the commodity. "Pitch farmers" who had leaseholds on Trinidad's Pitch Lake acted as suppliers, selling the lake asphalt to Barber's company, which laid down a finished product on the avenues for their clients back in the United States.

Staking out their claims around the approximately 100-acre lake, each farmer worked a leased holding. Digging down for a couple of days into the warm, claylike pitch, the farmers, their families, and hired hands worked in sync with the naturally recurring process, patiently waiting for the lake to replenish the just-excavated hole with new asphalt before starting the process anew. This was working well until the lake's landlord, the British colonial government of Trinidad, began drawing heavily on the natural deposit.

Proclaiming five acres of the lake to be theirs, the island's government was removing so much pitch so quickly that the entire lake was being drawn down. As this threatened the source and disrupted the supply cycle of the lake asphalt, the pitch farmers and Barber became alarmed and angered.

Having obtained a law degree from Columbian University (now George Washington University) in Washington, D.C., Barber led the farmers in a suit against the island's government. As Trinidad was under British rule, the case wended its way to the Privy Court in London, the highest court of law in the empire. In 1886, the Trinidad government won its case. As one paper of the day explained, "It seemed as though the asphalt industry was doomed to destruction. Mr. Barber then stepped into the breach and, acting unaided, carried on negotiations for two years with the island authorities. The result was that he undertook to buy out all the small pitch farmers and on Feb. 1, 1888, secured a concession for the entire Pitch Lake to the Trinidad Asphalt Company, which he had formed for the purpose. The concession was for twenty-one years, but soon afterward its term doubled." It was the most important deal in real estate Barber ever made.

Buying the lake out from under his former suppliers and securing a nearly half-century-long hold on the world's most productive source of asphalt put Barber in good stead. Five years later, however, Barber saw his monopoly threat-

Digging down for a couple of days into the warm, claylike pitch, the farmers, their families, and hired hands worked in sync with the naturally recurring process

ened again. This time it was by a rival firm called the Bermudez Asphalt Paving Company. When the Bermudez firm won a contract in Washington, D.C., a market Barber dominated, he retaliated with a lawsuit claiming that the Bermudez Asphalt Company was negligent and that their asphalt, pulled from an asphalt lake in Venezuela, was an "inferior material."

Creating a conglomerate of asphalt companies, he and a few wealthy industrialists established the Asphalt Company of

Creating a conglomerate of asphalt companies, [Barber] and a few wealthy industrialists established the Asphalt Company of America and gained control of the Bermudez company as a subsidiary.

America and gained control of the Bermudez company as a subsidiary. Barber and his new partners now had a leasehold on the Venezuelan asphalt lake, along with several other firms. As owners of his former competitor's supply of asphalt, just 100 miles from the Trinidad Asphalt Company's own pitch lake, Barber was no longer calling the Venezuelan lake asphalt an inferior material.

Amzi Lorenzo Barber was made the chairman of the Asphalt Company of America, which became commonly known as the Asphalt Trust. With interests stretching over three continents, with offices in London, Caracas, and New York, and with controlling interests in nearly 80 percent of the asphalt companies in the United States, he was at the peak of his career. Unfortunately for Barber, the fact that the company had borrowed $30 million from the Land Title and Trust Company of Philadelphia and then overpaid for the firms they purchased meant the Asphalt Trust was stretched to the brink. The upcoming Asphalt War was going to put them over the edge.

THE ASPHALT WAR

In 1900, when Barber's group got control of the Bermudez asphalt, it was hard to find an American who knew much about Venezuela. Just as difficult was finding a Venezuelan who knew much about the country's Bermudez region. The only human activities in the swampy area before the asphalt companies arrived were the aborigines waterproofing their canoes with the pitch from what the locals called Largo La Brea—the Lake of Pitch.

The Lake of Pitch in Bermudez was of paramount importance to the Asphalt Trust. Barber's success with similar Trinidad lake asphalt, and the world's growing demand for the asphalt pavements, turned the five-mile-long and three-mile-wide lake into an Alamo of sorts for the Asphalt Trust.

Taking control of the lake after purchasing the Bermudez Asphalt Paving Company, Barber and his partners were as motivated to harvest it as they were to keep rival companies from owning any part of it. Maintaining control over the natural resource was a make-or-break proposition to the fiscally stretched trust, as it allowed them to dictate the price per ton paid in the market as well as to help them secure lucrative paving contracts. One correspondent of the day claimed, "It is impossible to estimate its actual value," but ". . . it is a huge black sea of wealth."

It was less than a day's trip for the trust's supply freighter to the mouth of the

A lake full of pitch looks barren and lifeless but meant wealth and control to Barber.

San Juan River. About 30 miles up the muddy river, the ship docked at the company's modern 350-foot wharf at Guanaco. Not far away was a customhouse erected by a suspicious Venezuelan government for the sole purpose of verifying the multinationals' take from the lake.

The company's compound was a cluster of simple structures, connected by narrow-gauge rail lines and two rickety bridges to the asphalt lake. The only building of any significance at Guanaco was "the company's headquarters and store, built of concrete, with twelve-inch walls, and intended as a fortress in case of an attack." A conflict seemed imminent. Who was going to attack was not too clear—the Venezuelan Government's regular army, a group of revolutionaries that supposedly controlled the area, or even a New York syndicate or a local company. Everyone had a claim on the lake.

The reason for their interest was clear. One journalist noted, "The asphalt from the wonderful lake is the finest and purest in the world," claiming the raw material was 97 percent pure. The correspondent continued, "The refining process of this class of asphalt is simply heating it until the water is evaporated, for there is no scum, dirt, or foreign matter in it. The supply is practically inexhaustible, and should last until the end of the world or the bottom drops out of it."

A BISCUIT SALESMAN, A BANDIT, AND THE BATTLE FOR ASPHALT

A biscuit salesman named Horatio R. Hamilton, sent to Caracas by an American firm, had started a chain of events no one could have predicted. Not very good at selling his firm's biscuits but skilled at climbing social ladders, Hamilton married the niece of President Guzman Blanco. As dictator of Venezuela, Blanco granted Hamilton the rights to develop the Bermudez region however he could. In 1882, finding a New Yorker who wanted to get into the asphalt paving business, Hamilton struck a deal that brought him and President Blanco wealth while it gave the New York Bermudez Company control of the Lake of Pitch. In five short years, Hamilton spent his money, lost his job, and became penniless. He and his wife left Venezuela in shame.

Plundering Venezuela for what he could, Blanco allegedly embezzled $30 million of his countrymen's money—some of it profits from the sale of asphalt. He then took his spoils and went to Puerto Rico, leaving the door open for another crooked dictator.

General Cipriano Castro, the new president of Venezuela, was considered "a

Venezuelan
president
Cipriano Castro

bandit in the office." His slogan for the nation was "South America for South Americans." Citing the foreign interests of the Asphalt Trust, who now claimed control of the Lake of Pitch, he labeled Barber and his partners as greedy outsiders and warned them that they did not have control of the lake. Instead, he asserted that the new concessions he was granting to the Warner-Quinlan Syndicate of New York and a group composed of Venezuelan investors were legal. Barber and his partners didn't see it that way.

One of those partners was General Francis Vinton Greene, who from the beginning had been a central figure in the Barber Asphalt Paving Company. Having graduated at the top of his class at West Point, and having followed this success with a bright military career, he eventually became the District of Columbia's municipal engineer—a plum waiting to be plucked by the Asphalt Tycoon. Barber hired Greene as one of his vice presidents, and soon he was promoted to president of Barber Asphalt Company and then the president of the Asphalt Trust. In Venezuela, his West Point training and business acumen were put to use.

General Francis Vinton Greene

Seeing their interests in the Lake of Pitch threatened by President Castro, Barber and Greene chose to use force to protect their interests—they financed a band of revolutionaries to overthrow Castro. The band's self-proclaimed chief, General Manuel Matos, set out to topple the president. Generals Greene and Matos rendezvoused in Paris and discussed strategy. In agreement to form an alliance, General Greene departed for London and Glasgow, visiting shipyards in search of a vessel to smuggle weapons and ammunition into Venezuela.

The New York Times later reported that "as the results of Greene's work in Europe, the fatally celebrated steamer *Ban Righ,* whose history of piratic depredations is known to all, came to the coast of Venezuela, carrying an immense amount of munitions of war, and having on board the chief of the revolution [Matos]."

At the beginning of the 20th century, Americans were still excited about their victory in the Spanish American War. With a surge of internationalism, Americans were proud of having destroyed the Spanish fleet off the coast of Cuba, of occupying Puerto Rico, and of laying claim to the Philippines. President Castro knew he was in danger of antagonizing the Asphalt Trust and encouraging the United States to target Venezuela for an invasion.

With the help of a group of powerful U.S. Senators and the media, the

Barber and Greene chose to use force to protect their interests— they financed a band of revolutionaries to overthrow Castro.

Asphalt Trust worked hard at intimidating Castro and convincing the public that American interests were being threatened. The highly respected General Greene was walking the halls of the State Department garnering additional support.

On January 17, 1901, *The New York Times* carried a front-page story with the headline "Venezuelan Crisis Acute, Armed Force Demands Surrender of the Asphalt Lake, The *Scorpion* Ordered to the Scene to Prevent Dispossessions of Americans—To Use Force if Necessary."

> The property of the Asphalt Trust has been attacked by an armed force of Venezuelans. The trust has a private army of its own on the scene and the attacking force has ordered the Americans to lay down their arms. The *Scorpion* has been ordered to the scene and there is every likelihood that other American warships will be sent there. . . two ships seized by the Venezuelan Government and belonging to the Orinoco Shipping and Trading Company are believed to be about to proceed to the asphalt region as ships of war. They are the best ships in the Venezuelan Navy.

In reality, Venezuela seized the trading company's freighters because it was a country without any ships for a navy. Regardless, the revolution had begun, and over the next two years it claimed the lives of 12,000 men—more than the entire standing army of Venezuela. In the end, Castro remained in power and took control of the Lake of Pitch, and the Asphalt Trust went bust.

THE END OF THE ROAD

Barber lost control of his asphalt empire. He was forced out of the Asphalt Trust and resigned as its chairman. General Greene fared better, retaining his title as president of a new trust. In 1903, under the name of the General Asphalt Company, Greene and his new chairman absorbed 54 of the 69 working firms in the United States, putting them in control of nearly 80 percent of the market. The original Barber Asphalt Paving Company continued to exist as "the operating arm" of the new organization.

In testifying during an investigation into the Asphalt War and his involvement with it—a matter that the U.S. Congress, the Venezuelan government, and President Theodore Roosevelt were still trying to clear up years after the fact— Barber took the stand all day. Shielding General Greene, he said that he, not the general, had approved the funds for procuring arms to protect their investments

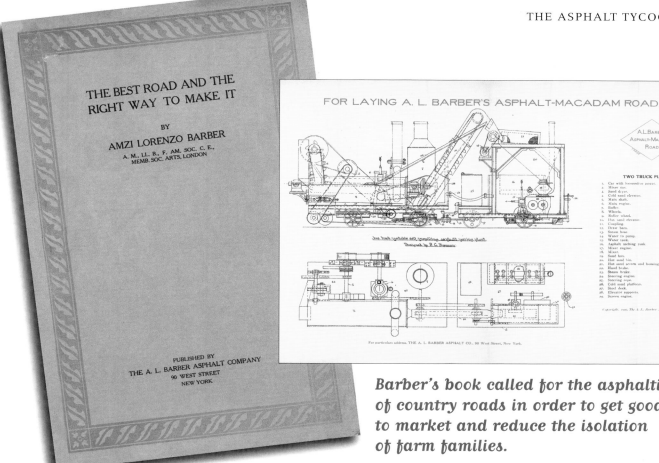

Barber's book called for the asphalting of country roads in order to get goods to market and reduce the isolation of farm families.

in Bermudez. In the end, the trust, after citing the struggles in Venezuela and increased competition in the U.S. as the key reasons for their inability to make timely payments on the $30 million loan, eventually was able clear up its financial obligations.

Stepping down from his chairmanship of the Asphalt Trust didn't mean quitting the asphalt business for Amzi Barber. He stayed active for several more years. In 1909, his book *The Best Road and the Right Way to Make It* was published under the aegis of his new firm, the A. L. Barber Asphalt Company. The book called for the asphalting of country roads in order to get goods to market and reduce the isolation of farm families. It promoted the idea of investing more federal money in the building of roads. Barber encouraged the asphalting of rural highways as a way of eliminating the hazards and annoyance of dust and mud. Not mincing words, the book's foreword heralded his achievements.

Only days after its publication, Barber, aged 66, was taken ill with pneumonia. On the night of April 17, 1909, he died at his palatial home in Ardsley, New York.

A vast expanse of asphalt bordered New York City's Madison Square Park in 1907.

CHAPTER SIX

"Artificial" Asphalt

"If we did not have asphalt, we would have had to invent it."

THE BARBER ASPHALT COMPANY conducted a survey while following nearly one million horses for more than six months. They then announced their findings: The average horse slipped and fell only once every 585 miles while traveling on asphalt, compared to a slip and fall every 413 miles on Belgium blocks or cobblestones. The more asphalt, the fewer slips, falls, and injuries—asphalt was good for you, and it was even better for your horse.

In 1893, proclaiming their results in pamphlets distibuted around American cities, the Barber Asphalt Paving Company also boasted that asphalt was more durable, smoother, quieter, and cheaper, even proclaiming asphalt pavement was "healthful" because its smooth surface was easy to clean compared to Belgium blocks and other stone pavements that collected horse urine and manure in their crevices. After scrutinizing nearly a million beasts, Barber's men had earned the right to make their equestrian proclamation.

By 1900, these facts were old news and the Asphalt Trust was the sole proprietor of the supply of nearly all the asphalt put down as pavement in America's major cities. The smaller cities and towns were unable to afford the desirable pavement. The only vessels delivering the Asphalt Trust's Trinidad and Venezuelan asphalts were oceangoing vessels, and the only practical place to put down an asphalt pavement was in a city that had access to a seaport. Otherwise, freight charges were cost-prohibitive.

Charles Duryea, now confident enough of his Buggyaut to drive it in daylight, poses in his groundbreaking gasoline-powered machine.

At this time, a revolution outside of the cities and over the open roads was unfolding with the work of two young men. Charles Duryea, 31 years old, and his little brother Frank, 23, had worked together to develop the first commercial gasoline automobile, starting with a broken-down horse carriage with wooden wheels that Charles had bought for $70. They rigged the dilapidated buggy with a crude one-cylinder gasoline engine and were running it up and down the streets of Springfield, Massachusetts, by 1893. They drove it at night so if the vehicle failed it wouldn't cause them embarrassment. Running it up against a curb was the only way to stop it. Producing more smoke than locomotion, it was called the Buggyaut—part buggy, part automobile. The odd-looking rig changed America forever, first creeping, then leaping onto the American scene.

At the turn of the century, people were moving about as never before in the history of the world. Baptized with dust or worse, swallowed by mudholes, these people—usually of means and influence—were witnessing for the first time the horrid conditions of the nation's rural highways. Collectively they used their power and began campaigning for good roads. They eventually, along with the farmer and the Middle American, called for local, state, and federal governments to improve their lives by improving the roads. The nation was nearly unanimous in wanting smooth and dustless roads.

LUST FOR DUSTLESS

In the United States before the 1900s, there was no such thing as a highway paved with asphalt running between two cities or states. They simply did not exist—not one. Some roads were gravel, others were compacted clay or crushed seashells, but none of them were asphalted all-weather roads, the kind that could be driven on no matter how much rain fell.

The country at the beginning of the 20th century was an expanse of dust, mud, or iron roads. The railroads had ruled transportation for more than half a century. If you dared to break from their steely rails and restrictive services and venture over the land, you were forced to deal with dusty, bumpy, and unmarked dirt trails.

Only in the largest cities would you find paved avenues; and even there, only the primary boulevards were paved. Some were paved with asphalt, some were brick, and others were built with wooden blocks. The secondary streets, however, were still dirty and often muddy. In the city, a pedestrian or horseman had a choice and could avoid the muddy ways by taking a main boulevard or sticking to wooden sidewalks. In the country, citizens were hostages to bad roads.

In his book *The Best Road and the Right Way to Make It*, Amzi Lorenzo Barber wrote, "Statistics in former years show that as a class farmer's wives provided a greater percentage of insanity than that of any other class. . . . Automobiles and good roads will furnish complete remedies for these two handicaps [isolation and insanity] to farm life." Promoting the construction of more and better roads was obviously in Barber's best interest financially, but his underlying point was indisputable: paving muddy roads would improve the lives of millions of Americans.

The country at the beginning of the 20th century was an expanse of dust, mud, or iron roads.

74

Dust suppressants such as the Barrett Company's Tarvia were used before true asphalt paving was widespread.

The Dust Nuisance Can Be Suppressed

The asphalt was packaged for its journey east in wooden barrels hand-made by men called coopers.

At first the bicyclists ventured out of the cities, followed a couple of decades later by the automobilists. When these groups began invading the rural highways, they were kicking up more dust than ever. Airborne particles of dust were a hazard to inhale and an irritant in even the mildest form. Dust killed crops, trees, and plants along the roads, depriving the farmer of income and shade and leading to erosion of the earthen roads. One automobile traveling over a mile-long dirt road each day dumped as much as one ton of dust a year on the agrarian's homestead and fields. These tons of dust cut into the farmer's profits, lowered his family's quality of life, affected their health, and were a regular annoyance. The dusty roads were bad before the automobile started using them, and now they were nearly intolerable.

Paving the road with asphalt was going to "get the farmer out of the mud," his goods to market, and his children to school. Asphalt was the undisputed choice of citizens, their governments, and road builders. The question remaining was: Will the asphalt be natural (like the asphalt from rocks or from lakes of pitch) or artificial (that is, manufactured)?

ASPHALT AS FINE WINE

In 1908 an asphalt distributor in New York, "John S. Lamson & Bro., Importers and Dealers," placed an ad showcasing its ability to deliver the very best. As if the firm's asphalts were fine wines, both domestic and imported, they were shipped in wooden casks and imported from Cuba, Egypt, Syria, and Trinidad, and as close to home as California and Texas.

As early as the 1890s, Bakersfield, California, was filled with the small oil refineries of wildcatters. Borrowing the process of distilling alcohol, the speculators, usually gambling with everything they owned, devised an asphalt version of a still that allowed them to separate kerosene, gasoline, and asphalt from the crude oil. The asphalt from Bakersfield's oils was especially suited for road paving, but only a handful of people understood its potential.

The asphalt was packaged for its journey east in wooden barrels handmade by men called coopers. The craft, taken from 14th-century European wineries, required the bending and fitting of wooden staves or slats. A skilled cooper could produce a three-foot-high barrel capable of holding about 50 gallons of heavy liquid asphalt in a short time. Swinging hammers and using saws, day in and day out, left many of the coopers with a right arm much larger than their left.

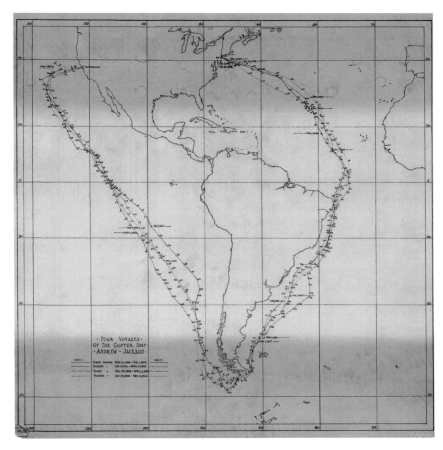

Before transcontinental railroads and the Panama Canal, asphalt shipped from California to the East Coast had to take the long route south, around perilous Cape Horn, and north again past all of South America.

Since aging was not an issue and shipping the asphalt in a fine oak barrel would be a waste of good wood, coopers were instructed to produce barrels out of less expensive woods such as pine. As soon as the heated liquid asphalt was poured into the wooden barrels, it was stored in a cooling warehouse. After cooling and solidification, the asphalt was shipped.

The Union Oil Company, established in 1890, was one of the first to send its asphalt back to the booming East Coast markets. Pennsylvania oils were geographically more available to the eastern markets but they had one shortcoming: Pennsylvania petroleum lacked the physical makeup for a high-quality asphalt. California's bitumen, on the other hand, was ideal for making asphalt.

Contracting with steamship owners, Union Oil sent its product "around the Horn." In the days before the Panama Canal, the most practical path by ship to the East Coast was through treacherous waters off the tip of South America. In the late 1800s, it was cheaper to make the long trip east by ship, but it wasn't long before the 50-gallon wooden barrels and circuitous routes of steamships gave way to beelining railroad tank cars holding thousands of gallons of oil.

Cutting into the Asphalt Trust's dominance of asphalt products, Union Oil proudly claimed that prior to 1901 and the advent of California petroleum asphalt, the cost of asphalt was $3.36 per square yard. After their entry into the market, the price per yard of asphalt was more affordable at $1.52 a yard. The asphalt industry's rich history of natural lake asphalts was legendary. Its bright future, however, was in "artificial" asphalts. Refined asphalts would steadily meet and eventually exceed the production of natural lake asphalts in the first 20 years of the 20th century.

As refineries tried to keep pace with the demands of the new automobile industry by producing more and more gasoline and oil-based lubricants, they found themselves left with more and more of a by-product called asphalt. Residing at the bottom of the barrels of crude oil after the removal of its valuable gasoline, the asphalt was at first considered waste.

In a self-fulfilling cycle, the more desirable the automobile became, the more fuel was consumed. With automobile enthusiasts topping off the tanks of their dream machines and with their collective voice becoming louder in demanding

The Union Oil
Company at Bakersfield, California, in 1910. From oil, it refined gasoline for the increasingly popular autos and asphalt for the roads they drove on.

paved roads, asphalt was no longer seen as an unwanted leftover, but rather as a profitable commodity. The economies of scale of the expanding petroleum industry meant the price of petroleum-based asphalt was falling as the demand for it was increasing, further fueling demand and creating a new market for artificial asphalts as a road-paving material.

ARTIFICIAL ASPHALT COMES OF AGE

Lake asphalts from Trinidad and Venezuela were "natural asphalt" and were therefore, according to Barber and his partners in the Asphalt Trust, good. Refined asphalts were "artificial" and therefore bad, they claimed. Natural was good, artificial was bad. It was that simple—or at least Barber and the Asphalt Trust's propaganda machine worked to make everyone believe that was so. Labeling asphalt refined from crude oil as artificial was their ploy to discredit the competition.

Lake asphalt may have been natural, but it was not without its issues. Proving itself as a reliable product since its American debut in Newark, New Jersey, in 1870, it was known for creating a solid and durable travel surface. Retrieving asphalt from the lake, however, was an expensive endeavor. By the time pitch

farmers mucked out the asphalt from the lake bed, dumped it into a railcar, hauled it to a refinery, packaged it in wooden barrels, and shipped it 2000 miles to the United States, it was more expensive than the so-called artificial asphalt. Lake asphalt, being a hard asphalt, needed to be softened with pricey liquids called fluxes. Compounding matters, the Asphalt Trust was known to price-gouge wherever and whenever possible. Where the Trust had no competition, the price per ton could be double the going rate of the artificial asphalts. Artificial asphalts eventually led to a reduction in artificial pricing.

Proof that lake asphalt was indeed natural was in the sticks, stones, and vegetation sometimes found in it. On the other hand, artificial asphalt was produced from oil out of a well and was for the most part free of unwanted natural ingredients. More significant than its purity was its abundance. An expert on petroleum-based asphalts estimated that 99.5 percent of the available asphalt from refineries in 1900 was going unused. If refineries weren't throwing this bottom-of-the-barrel residue away, they were giving it away for seven cents a gallon.

The oil refinery of the 1800s was a crude and sloppy facility that created a niche industry for locals. Little girls nicknamed oil dippers worked the rivers downstream from the rustic refineries. Wading in slow-moving water in their Victorian-era gowns, the oil dippers wielded wooden pails and skimmed from the water's surface any oil that had floated down from the refinery, depositing the results into even larger wooden barrels. Small groups of girls could collect enough oil by the end of a good day to make a tidy profit for their families by selling the "lost" oil back to the refinery that had spilled it.

Wildcatters—speculators drilling for oil and usually risking everything to do so—pulled whatever oil they could from the earth in a messy, inefficient, and wasteful process. Their simple goal was the distillation of oil to produce kerosene in hopes of selling the fuel for use in

Opposite:
Spindletop, the world's first oil gusher, blew in just south of Beaumont, Texas in 1901. Prior to this time, a productive oil well was one that produced fifty barrels a day. With a flow of over eighty thousand gallons a day, Spindletop ushered in the beginning of the modern petroleum industry. Wildcatters rushed to put their own wells alongside the first one, and a boomtown was born.

Boiler Avenue
April 23 1903
Edgerton

lamps and home heating units. With the exception of oil used as lubricants for steam engines and asphalt for a waterproofing agent, the best use for the remaining thick, heavy black bitumen in the bottoms of their barrels was for paving roads. Over time, the asphalt derivative would be recognized as the best and purest asphalt in the world. Still, Barber and the backers of natural lake asphalts persisted in calling refined asphalt artificial. At this time, automobiles were still only a novelty, a market for gasoline was just starting to percolate, and the oil industry was far from its full potential.

RUMORS AND REALITY

As early as 1884, the rumors about the discovery of natural asphalts in California were circulating around the East Coast. Jesse Warren, a superintendent of the Barber Asphalt Paving Company, was sent on a mission to the West Coast to confirm these reports and, more important, to determine whether there was enough asphalt in California to upset Barber Asphalt Paving Company's monopoly. If there was, the second part of Jesse Warren's mission was to see if the firm should expand its operations to prevent California's natural asphalts from "making inroads into asphalt paving business in the East."

His report explained, "[T]here is nothing either to fear in the East or warrant organizing a paving business in California . . . There was no asphalt in California other than the dunes of oil sand along the shore at Carpinteria, and harder grades of asphaltic rock at La Patera, Santa Cruz, etc., with some trickling of oily asphalt from those deposits and some oil springs and surface croppings near Bakersfield." Jesse Warren's report was based on his observations about the natural asphalts he had been sent to find. It ignored the potential of asphalts refined from petroleum, which was understandable given the times. Asphalt from crude was practically nonexistent, and the Barber Asphalt Paving Company was geared for the production and distribution of natural asphalts. With the exception of Union Oil, so too was the rest of the asphalt industry. To some, the idea of making suitable asphalt from crude oil was on a par with practicing witchcraft and selling snake oil.

To some, the idea of making suitable asphalt from crude oil was on a par with practicing witchcraft and selling snake oil.

In 1898, Jesse Warren's brother Fred, who also worked for Barber Asphalt, made his own trip out to California to investigate the oil industry. True to brother Jesse's report more than a decade earlier, Fred found that not much was happening with the West Coast asphalt industry. The Alcatraz Asphalt Company was struggling to come up with a working substitute for Trinidad asphalt by working

The California coast had natural asphalt, which the Barber companies considered desirable, as in this asphalt deposit. The Warren family realized the greater potential of manufacturing asphalt from the abundant oil found in the state.

the "dunes of oil" and mixing the substance with rock asphalt from La Patera. The company's effort to build an expensive pipeline to move "liquefied bitumen" from the well to a refinery near a rail line was a bust. In Bakersfield, a real-estate turned oil-speculation firm was having better luck skimming liquid asphalts off natural pools of oil.

Still, Fred Warren suspected something that his brother and all the others missed in their pursuit of topical natural asphalts.

Fred Warren quietly instructed that a few barrels of the California oil be placed on his train back to Boston for a closer examination. The laboratory results proved his instincts were correct. As a reporter wrote years later, "He discovered what no one else had dreamed, that, if intelligently handled, the residue from distilling of California oil was the purest and best asphalt the world had ever known . . . a discovery that was to completely revolutionize the oil refinery and asphalt productions of the world."

THE WARRENS' ASPHALT DYNASTY

The Warren family assemblage was a historic event and a near-religious experience for the kin that faithfully gathered in 1899. Appropriately, the event was recorded in the family Bible. Albert, Fred, George, Herbert, Henry, Ralph, and

Walter called a meeting at the family's campground on Lake Ontario in Oswego, New York. The seven Warren brothers asked their two sisters, Ella and Mabel, to join them, as well as the family of recently deceased brother Jesse. The convocation lasted a few days. The gravity of the situation was heavy. The Warren children, who were to be seen and not heard, whispered questions to their mothers about the purpose of the assembly.

The sole reason for the gathering was to answer one question: Should the Warren family organize its own business, the Warren Brothers Company? Tempting as it was to strike out on their own, doing so would be challenging Barber and his newly formed and powerful Asphalt Trust of which the Warren family had just become a part. In a bizarre twist, while Jesse was living, the Barber Asphalt Paving Company was employing one half of the Warren brothers. In addition, the men, women, and children of the Warren family were stakeholders in the Asphalt Trust through a complex web of stock holdings and equity positions.

The question of starting a new company, independent from Barber's conglomeration, was a difficult one. An industrious and ingenious lot, the Warrens were game for the challenge, having proved themselves many times before. Conservative and loyal, they found it hard, however, to make a break from such a tried-and-true partnership as the one they had with Barber. Besides, much was at stake—the family's reputation, its financial resources, and its industrial legacy. Extending as far back as the brothers' grandfather, another Jesse Warren, the clan's resourcefulness was indisputable.

In 1837, grandfather Jesse Warren, an inventive blacksmith, had set up business in the rustic and rural area of Peru, Vermont. There he manufactured his "Side Hill" plow with his oldest son, Joseph. For over a hundred years, this plow helped many agrarians in the mountains of New England till their fields. When a fire burned his business to the ground, the Warrens relocated and rebuilt their works in Glenn, New York, and continued manufacturing the family's products.

In 1847, five of Jesse's six sons struck out into businesses of their own. Heading to Ohio, they pursued profits in roofing materials. Jesse's oldest son remained behind to help him with his Side Hill plow.

Within a few years, the brothers were shaking up the roofing industry by introducing new methods and materials. John, the oldest of the Warrens in the business, became the first to distill coal tar, a residue of waste from the process of burning coal for energy, and turn it into a useful pitch for waterproofing. The industry shifted its focus from the distillation of pinewood and pine pitches to

The question of starting a new company, independent from Barber's conglomeration, was a difficult one. An industrious and ingenious lot, the Warrens were game for the challenge.

that of coal tar. The Warrens were leading the industry's charge, opening offices in Boston, Buffalo, Louisville, New York, Philadelphia, and St. Louis. By the mid-1800s, the Warrens' companies were the largest consumers of coal tar in the United States.

With gasworks around the country practically giving away their coal-tar waste, the Warrens expanded their dealings to New York City. Striking a deal with the city for the coal-tar waste from its 42nd Street Gas Works plant, the Warrens entered into a 30-year contract to remove each barrel of the byproduct for just 50 cents. When it was discovered that coal-tar-naphtha enhanced the dyes used in the arts, the family "gained large profits by selling it."

Another Warren brother, Cyrus, after five years in the roofing business, left Ohio and headed back to the East Coast, where he studied chemistry at Harvard University. Taking his brother's discovery of distilling coal tar a step further, Cyrus invented an oil-refining process known as fractional distillation. Shortly after his discovery, he briefly held a professorship in organic chemistry at the Massachusetts Institute of Technology, but it "consumed too much of his time," according to a family member. Profits, not professorships, were to be his life's work. Regardless, Cyrus Warren became known as "the best authority on the use of bituminous materials."

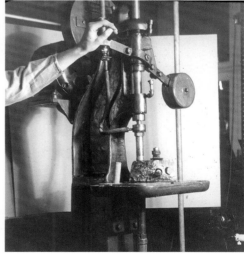

Warren laboratories like the one shown left were state-of-the art. Above, a sample of hot-mix paving material is tested for stability by punching it. Such a test made sense at a time when steel wheels on trucks were common.

Having the best practical experience and the set of skills closest to what was needed in the upstart petroleum industry, the Warrens set up a large oil refinery in New York City, becoming one of the first to handle the distillation of Pennsylvania crude oil. Unfortunately, Pennsylvania's oil was lacking an asphaltic base and was disqualified as a source for artificial asphalt.

More devastating than the fire that had wiped out grandfather Jesse's foundry was a blaze that destroyed the New York refinery in 1869. Of the five brothers who originally owned the facility, only John—who had been the first to distill coal tar—decided to rebuild. Going it alone, he had begun the reconstruction process when he was made an offer he chose not to refuse. Selling the partially completed refinery to the Standard Oil Company, John also made a break from the oil-refining business.

At this time, the youngest of Jesse's six sons, Burgess, was in the roofing business in Washington, D.C. For a solid decade, the District of Columbia was using large amounts of coal-tar cement and Burgess was busy selling it to them in varying forms mixed with stone, sand, lime, or even sawdust. Eventually, the Warrens were not alone in the business of coal-tar refining, and competitors were driving the profit out of the business. With their long-term contracts for purchasing coal tar from municipal gasworks expiring, Burgess and the other Warrens turned their attention to a new product, Trinidad lake asphalt.

Impressed with the lake asphalt's performance on Pennsylvania Avenue in 1876, the Warrens began investigating its source. It wasn't long before they aligned themselves with Barber and his company. Burgess, the designated point man for the Warren clan, invested much of the family's resources in the partnership with the Barber Asphalt Paving Company, including the Warren Chemical & Manufacturing Company and then a later firm they called the Warren-Scharf Asphalt Paving Company. Supporting Barber's efforts in obtaining a long-term lease on Trinidad's asphalt lake positioned the Warrens to profit from Barber's success with the natural asphalt. The Asphalt Tycoon and the Warren Dynasty were united . . . for a while.

Tragically, while conducting the family's business in 1880, one of the Warren brothers was lost. Traveling between his home in Boston and New York City, 53-year-old Herbert Warren, the father of 10 children between the ages of six and 28, was killed when the steamship *Narragansett* collided with another vessel in Long Island Sound. In the Warren tradition of carrying on after a hardship, the older children left school to help support their younger siblings, joining their uncles Cyrus and Burgess in the family businesses. A few years after the catastrophe, the Warren-Scharf Asphalt Paving Company was up and running and the estate of Herbert Warren was among its largest shareholders.

Fred came up with an idea while under the employment of the Barber Asphalt Company, and this idea led to the family meeting of 1899. The Warren family had profited greatly while partnering with Barber—enjoying the benefits of their cartel's dominance in selling and laying Trinidad's natural lake asphalt. It was the Warren family's determination to stay at the vanguard of the road-building industry, however, that led them to their next defining moment. At the four-day family retreat in upstate New York, the grandchildren of Jesse Warren, now into their 20s and 30s and with children of their own, decided to establish the Warren Brothers Company. The new venture would make the Warren name synonymous with asphalt for the next half century.

Facing page:

The Warren family portrait shows the seven founding brothers together with brother Jesse (deceased before the company was organized) and two sisters. The photo was taken in 1886. Seated from left to right are Walter B., Ella, Albert C., Mabel, and Ralph L. Warren. Standing from left to right are Herbert M., Frederick J. (first President of Warren Brothers Company, 1900–1905), Jesse, Henry J., and George C. Warren.

Cross section of a Bitulithic type mix, patented in 1901 by Frederick Warren.

Most hot asphalt mixes of the day were called sheet asphalt and used sand instead of larger stones. Bitulithic Macadam was laid in a two-inch-thick surface over old macadam pavements.

BITULITHIC MACADAM

On June 4, 1901, the Warren family received good news. The United States Patent Office recognized the family's customized Bitulithic Macadam, a hot-mix asphalt. The patent supported the Warrens' legal efforts to protect themselves from others who might copy their formula and attempt to profit from it. Bitulithic Macadam's name reflected a mix that was part bitumen (asphalt) and part lithos—Greek for stone—and relied on the principles of John Loudon McAdam: a structure with interlocking stones that was strong enough to support the traffic's weight but was flexible and resisted breaking up. It was also the first asphalt mix blended with scientifically proven consistency. The same year the patent was issued, Park Place in Pawtucket, Rhode Island, was paved, becoming the first road to be surfaced with the material.

Most hot asphalt mixes of the day were called sheet asphalt and used sand instead of larger stones. Bitulithic Macadam was laid in a two-inch-thick surface over old macadam pavements. If the road was being constructed from scratch, a three-inch layer of Bitulithic Macadam was recommended. The simple rule regarding aggregate was that the largest stones could be equal to half the depth of the road. Thus a three-inch-thick roadway of Bitulithic Macadam could include stones up to one and a half inches in diameter. (In the 21st century, guidelines suggest a ratio of three to one for fine mixes and four to one for coarse mixes.)

Using 60 percent more stone than sheet asphalt, and using larger stones than previously considered possible, the recipe was a breakthrough. Called an asphaltic concrete because of its use of larger stones, Bitulithic Macadam was made up of large, medium, and small stones. The largest stones acted as the pavement's backbone, giving it strength for heavy traffic loads. The medium and smaller stones made it possible for all of the aggregate in the mix—large, medium, and small stones—to lock together in a compact and therefore dense mass of stone and asphalt. This density was the key to the mix's success. Previous mixes using large stones had problems with large voids. It was expensive to fill large pockets of air with asphalt. With Bitulithic Macadam, the density of the aggregates resulted in an economical but strong and flexible road structure.

The Warrens, eager to create an air of mystery around their product in hopes of discouraging other contractors and municipalities from counterfeiting their recipe, emphasized the scientific nature of their work. One example was the paper presented to the Boston Society of Engineers in 1914 by the Warren Brothers Company's superintendent of refineries, who claimed, "It is impossible

for us to formulate a standard screen test or rule as to just what proportion of each size of particle of mineral aggregate is required to produce maximum density or minimum voids because the stone from different quarries crush to different shapes, and for this reason it is necessary for us to maintain a corps of laboratory representatives who travel continually from plant to plant setting mixtures to fit requirements . . . by careful laboratory test."

At first the family faced the dilemma of success: How big a territory should Warren Brothers Company carve out for itself? Overexpanding could be as disastrous as not being ambitious enough. Focusing at first on New England and upstate New York, the Warren Company eventually had offices in Dallas, Houston, Nashville, New York City, Portland, St. Louis, and even Toronto, Canada. So quick was their expansion that Fred Warren was on the road constantly, establishing subsidiaries, promoting his patented product, and defending it in court. In 1905, exhausted from a particularly strenuous trip, he phoned his office explaining that he would not be in as scheduled and was instead checking into a hotel room to rest. That night, at the age of 39, he died of a heart attack in his room. His original patent for Bitulithic Macadam, however, lived on until 1918, and various forms of the patent carried over into the 1930s.

The family faced the dilemma of success: How big a territory should the Warren Brothers Company carve out for itself?

PORTLAND, MAINE, HIGH STREET, PAVED WITH BITULITHIC.

SPRINGFIELD, MASS., MAPLE STREET, PAVED WITH BITULITHIC.

NORWICH, NEW YORK, HENRY STREET, PAVED WITH BITULITHIC.

BITULITHIC THE WORLD'S MODEL STREET PAVEMENT
ITS WONDERFUL GROWTH DUE TO MERIT ALONE

AN UNRIVALLED RECORD.

7 cities in 1901 laid	. . . .	16,400	square yards
33 " " 1902 "	. . . .	400,831	" "
40 " " 1903 "	. . . .	915,630	" "
45 " " 1904 "	. . . .	940,239	" "
63 " " 1905 "	. . . .	1,091,825	" "
79 " " 1906 "	. . . .	1,508,093	" "
67 " " 1907 to June 25, have laid or contracted for		2,105,397	" "

Total in six years 6,660,612 square yards, equal to 378 miles of 30-foot roadway.

WARREN BROTHERS COMPANY

93 FEDERAL STREET - - - BOSTON, MASS.

Registered Trade Marks

"BITULITHIC"   "BITROCK"   "PURITAN"
"BITUSTONE"   "BITUMINOUS MACADAM"

CINCINNATI, OHIO, FAIRFAX AVENUE, PAVED WITH BITULITHIC.

PATENTS, PROMISES, AND COURTROOMS

Succeeding Fred at the Warren Brothers Company was brother George. Defending the family's patents had become central to the company's profitability. Speaking at Columbia University to a group of highway engineering students 10 years after his brother's death, George Warren explained the principles of patents in American history. Citing the first patent in the United States, awarded July 31, 1790, to Samuel Hopkins of Vermont, George noted that the patent's paperwork was important enough to bear the signatures of George Washington and Thomas Jefferson. He also pointed out that the Founding Fathers, backed by the Supreme Court, had proclaimed patents as central to the fostering of innovation and to the building of a robust economy.

Patents in pavement were not new—Nathan B. Abbott of Brooklyn, New York, had filed the first one in 1871, explaining, "I have invented certain new and useful improvements in Composition Pavements; this invention relates to that class of composition pavement composed mainly of sand, gravel, ashes, broken stone, etc., mixed with tar or other bituminous material to form an elastic and water proof road or pavement." One year later he updated his patent, effectively improving and extending it. He explained the benefits of heat: "Instead of mixing with the same sand, gravel, ashes or other material, I employ the bituminous material in a hot state, and spread the same thoroughly and evenly upon the pavement surface to be coated by means of a mop or broom." Thirty years later the Warren Brothers were granted their own patent that rearranged the quantities of sand, increased the gravel, and omitted the ash. They would name it Bitulithic Macadam.

George Warren spoke from personal experience when it came to enjoying the monopoly provided to patent holders as well as the rights that went along with a patent. George quoted, with heavy emphasis, the words from a Supreme Court Justice in the patent case of *Kendall* v. *Winsor,* that patent holders were "entitled to protection against frauds or wrongs practiced to pirate from them the results of thought and labor in which nearly a lifetime may have been exhausted." George lectured that municipalities that avoided using patented products would fall hopelessly behind the times—17 years behind the times, the life of a patent. Many cities disagreed and saw patented products as risky and legally problematic.

The first attack on Bitulithic's patent rights was from the Warrens' old business partner and new competitor: Barber. The Asphalt Trust filed suit in an attempt to block Bitulithic's patent by declaring it invalid. The trust hoped that

Opposite:

The Warren family was aggressive in promoting its patented Bitulitic pavement. Medallions were placed in each project, and postcards were sent to contractors showing completed jobs.

The first attack on Bitulithic's patent rights was from the Warrens' old business partner and new competitor: Barber.

When the Warren Brothers repaved this road, they responded poetically to complaints about its former condition.

the Warrens would be discouraged and abandon their plans to promote it. Bitulithic threatened the Asphalt Tycoon's world as it used only about half as much asphalt as Barber's sheet asphalt. Barber knew better than anyone that Bitulithic's success would dramatically reduce the tonnage of Trinidad asphalt used nationally. When the Warren Brothers Company's patent was upheld in court, Barber continued his attempt at discrediting Bitulithic and its supporters within the paving industry. At first he was successful, but eventually his plan backfired. As an industry specialist explained, the Warren Brothers Company was "happy to have it be general knowledge that large stone was too difficult for mere humans to lay."

When Barber and the Asphalt Trust cut off the supply of natural asphalt to the Warrens in an attempt to starve them of Bitulithic's crucial ingredient, bitumen, the family fell back on their expertise in coal tars. Anticipating Barber's move, the first Bitulithic plant was built immediately next to the Cambridge Gas Works in East Cambridge, Massachusetts, so it could mix the first batches of Bitulithic with coal tar. Later, the Warren Brothers Company entered into a pact with Standard Oil to produce artificial asphalt, marking the beginning of the end of Barber's and Trinidad lake asphalt's reign and the beginning of artificial asphalt's dominance. In 1900, nearly all asphalt in the nation was natural Trinidad or Venezuelan asphalt. By 1907, imports of natural asphalt had dropped by 80 percent, from a high of 320,416 tons to just 65,000 tons.

Hand in glove with the patents were warranties. Municipalities were eager for asphalt pavements, and by the turn of the century, New York City was requiring that asphalt be used instead of bricks, granite, and wooden blocks; but cities wanted a guarantee that patented asphalts would work and sought warranties on paved roads for as long as 15 years.

One obstacle that patented asphalt mixes created for awarding authorities was keeping the competitive-bid process alive while using a legally protected product. To allow Bitulithic and other patented pavements into the low-bid system and to protect themselves from sudden increases on tightly controlled and patented products, municipalities began demanding license-mixture agreements. The Warrens and others drew up legal paperwork promising that the city or its contractor was entitled to purchase their asphalt mixes at a predetermined price per ton.

The engineering of mixes was still in its infancy. The earliest test was the chew test. A foreman checked the quality and properties of a given asphalt cement by placing a piece of the material in his mouth and chomping on it to get a sense of how flexible or rigid it was. Based on the foreman's impression of the asphalt's properties, more or less additives were placed into the melting kettles

By 1913, improved asphalt paving was welcomed by the wealthy residents of Fifth Avenue mansions. The 1869 version had created black dust that coated everything within and outside their homes.

of bitumen. For a hard asphalt, fluxes were added; for softer materials, more bitumen was added. When a penetration test was developed, many foremen challenged progress, claiming that their chewing often predicted the chemical makeup and elasticity of bitumen as well as or better than the newly developed apparatus.

In hopes of minimizing their exposure to risk, paving companies started conducting scientific research on how to make their pavements last longer. It was becoming key to any paving company's longevity. Heavy steel wheels and solid rubber tires on the vehicles of the day were starting to rip apart previously suitable pavements. It now behooved contractors to understand their hot-asphalt mixes better than ever before. Whether they were learning how to use steamrollers to press the air voids out of pavements or increasing quality controls for stone and bitumen, road-paving contractors were working harder to make their roads more durable to avoid costly rebuilds.

In reality, the warranties of that period were often worthless and always costly. Anticipating repaving, contractors raised the price of laying asphalt. When pavements failed, many contractors went bankrupt trying to fulfill their warranties. Cities, on the other hand, were often left with empty promises and empty guarantees to repave rutted streets. Worse for the taxpayers but good for the Warren Brothers Company were the failures of businesses that made guarantees but went bankrupt by the time their surfaces failed. As a result, proven pavements like Bitulithic and enduring firms like the Warren Brothers Company were in high demand.

By using less asphalt and more stone, contractors willing to infringe on the Bitulithic patent could fatten their wallets.

THE TOPEKA DECREE

To many road builders, the Bitulithic model was a road map to increased profits. By using less asphalt and more stone, contractors willing to infringe on the Bitulithic patent could fatten their wallets. Many contractors began mixing large local stones into their asphalt mixes while others began using stones instead of just sand in their sheet asphalt mixes and minimizing the use of expensive bitumen, regardless of their clearly violating the Warren Brothers Company's legal claim to its revolutionary use of aggregate in an asphalt mix. Adding to the Warrens' concern, widespread and reckless applications of aggregate in asphalt threatened the family's reputation and the credibility of Bitulithic.

In an effort to curb the abuse, George Warren filed suit in Topeka, Kansas, against Kaw Paving Company, a violator of the Bitulithic patent. The Warren

Brothers Company had a strong legal department, and judges often sided with the company's arguments and efforts to protect their patents. However, having endured expensive legal battles with the likes of Barber, George decided it was wiser to make an out-of-court settlement. Signaling an accord with the rest of the asphalt paving companies, the Topeka Decree was put forward on May 26, 1910.

Specifically, the Topeka Decree laid out guidelines between the former defendant, Kaw Paving Company, and the plaintiff, the Warren Brothers Company. The announcement ensured that Kaw Paving and other contractors would not be breaching the Bitulithic patent as long as their asphalt mixes were made of stones smaller than one half inch in size. Most of the stones in these mixes, 90 percent of them, had to pass through a screen with openings of one-quarter inch.

With the Topeka Decree removing the threat of legal action against imitators of Bitulithic, many new fine-graded mixes hit the market. Using mixes that didn't have Bitulithic's larger stones, a contractor's employees were able to mix and shovel the small-stone mixes faster and with greater ease, cutting down on the cost of labor. With more stones and less asphalt, the contractors were able to further reduce their costs. The Warrens agreed to the decree because it clarified what was rightfully theirs, a large-stone mix with great strength.

The Warrens knew that contractors were adding stones to cut cost, not to increase the quality of the asphalt mix. They felt that stones smaller than a quarter of an inch didn't give the asphalt enough strength to compete with Bitulithic. The Warrens went as far as saying that the small stones the contractors were forced into using were like "raisins in a fruit cake," the analogy being that better and larger fruits in a cake were like the stones found in Bitulithic. Bitulithic became so popular that its name became interchangeable with the word asphalt. Confusing matters even more was the long list of other patented mixes such as Interstate Amiesite, Wilite, Romanite, National Pavement, Imperial, Indurite, and Macasphalt.

Almost 100 years later, experts in the paving industry would still lament the Topeka Decree, citing the genius in Bitulithic's large-stone aggregate.

ARTIFICIAL ASPHALT'S REAL SUCCESS

Asphalt was better than the rest of the paving options. Foreshadowing the use of asphalt over the nation's highways, a mid-1920s survey showed that bituminous surfaces were dominating all other forms of pavement in the nation. More than half of all paved surfaces in metropolitan areas were asphalt—54.6 percent, to be exact. The sum total of all other surfaces combined did not add up to asphalt's impressive statistics: stone-block pavements lagged at 14.9 percent, traditional macadam at 13.3 percent, clay bricks at 12.8 percent, and concrete at just 2.7 percent. Last, but still in the game, wooden blocks made up 1.7 percent of the paved surfaces.

Municipalities often chose smooth asphalt over rigid and bumpy stone surfaces because many of their citizens demanded it. Up until the 1920s, human locomotion was still dependent on the horse, and both horses and their owners preferred the smoothness and forgivingness of an asphalt surface. Taxpayers whose property abutted an improved road and had to pay for paving improvements through assessments levied on them were choosing asphalt over all other pavement surfaces for the peace and quiet it promised them. When it came to making improvements or repairs to a city's subterranean infrastructure, asphalt proved the most workable material of all. Ripping it up and putting it back down was simpler than with any other material. With the gasoline-powered vehicles driving up the consumption of oil, plenty of petroleum was available for the production of asphalt pavements just as demand was increasing.,

On July 11, 1916, President Woodrow Wilson signed the Federal Aid Road Act of 1916, launching the greatest partnership in road-building history. For the first time in American history, the federal government and all of the nation's state governments entered into a nationwide joint venture to build a network of interstate highways, the likes of which the world had never seen. Bolstering research and development, promising large sums of funding and a comprehensive plan, the Federal Aid Road Act of 1916 ensured that the demand for asphalt would reach new heights. Only a world war could threaten the progress of the new road-building campaign.

Mobilizing the country for World War I deepened the nation's understanding of just how important roads had become to the economy and defensive readiness of the nation. Military mobilization efforts, however, nearly destroyed every road in the country. Behemoth trucks and their solid steel wheels ripped apart paved surfaces as they hauled war goods, dry goods, and anything the railroad companies couldn't handle on their overburdened freight cars. This wiped out many of the gains made during the years leading up to the war. Not until years after World War I would the rebuilding and improving of America's highway system get underway.

The chief agents for artificial asphalt's success were the Barber Asphalt

THE FEDERAL AID BILL

Senator Bankhead, of Alabama.

ITS MAKERS

Congressman Shackleford, of Missouri

President Wilson, the Final Signer.

Photos copyrighted by Clinedinst and Harris & Ewing, Washington.

On July 11, 1916, President Woodrow Wilson signed the Federal Aid Road Act of 1916, launching the greatest partnership in road-building history.

Paving Company and the Warren Brothers Company. In pursuit of their own interests, these two dynamic firms nurtured asphalt from its infancy to a mature industrial product. Acting as sponsors for the advancement of asphalt, Barber and the Warrens tested, applied, and promoted their products and in doing so advanced the use of asphalt as a pavement for the world. After the Great War, asphalt played a starring role in road building's Golden Age—the covering of the nation's highways with petroleum-based asphalts.

A Hetherington & Berner railroad asphalt plant, circa 1908, traveled from town to town, bringing the promise of smooth pavements.

CHAPTER SEVEN

Hot Mixing, Raking, and Rolling

"By and large, the difference between a rich nation and a poor one is the difference between a mechanized and a muscular society."

IN THE 50 YEARS between the first patented bituminous hot mix in 1871 and the last days of the original Bitulithic patent in 1920, the asphalt industry moved slowly through a long, awkward adolescence. The golden age of road building that was kicked off during the Roaring Twenties marked the industry's entrance to adulthood.

If not for the intrepid inventors, scientists, and businessmen who kept the mechanization process of the industry moving forward, asphalt might have died at birth. The development of the asphalt plants and the constant improvements to equipment and machines used to haul hot mix to the work site and lay it down made asphalt the material of choice for a new generation of road builders. But it was a childhood of hard knocks—an upbringing of trial and error.

Innovations during these first 50 years positioned asphalt for greatness. By the end of World War I, the maturing asphalt industry was taking its position at the forefront of the most ambitious road-building campaign on earth. Unfolding during what has become known as the American Century, the golden age of road building was under way. Three quarters of a century later, asphalt would provide the surface for 93 percent of all roads in the country.

97

In 1849, M. Merian, a Swiss-born engineer, was the first to create an asphalt roadway when he built a macadamized road of rock asphalt in Val de Travers, Neuchatel, Switzerland. The first use of special equipment for asphalt road paving appears to have been in Paris, France, in the 1850s, when the first asphalt contractors were applying the earliest asphalt pavements to the city's grandest avenues. Their unsophisticated methods and simple tools of the trade were quickly copied and used throughout Europe.

These early contractors placed chunks of rock asphalt into metal drums or heated the asphalt-impregnated aggregates on metal plates and baked them over open fires. Several hours later, the brittle rock asphalt was crushed into a powder, and workers sifting through the warm dust removed lumps of stone or asphalt that might have caused unevenness in the pavement. Wasting nothing, they pounded this lumpy debris further and either dumped it back into the current batch or saved it for the following one.

Hauling the rock and asphalt down the cobblestoned streets of Paris in small pushcarts or wagons, the crew dumped the now cold mix onto the street. Workers with heated rakes moved the rock asphalt about, spreading it to the desired thickness. Following the rakers were men with heated tampers that gently tamped the asphalt down, compressing it but keeping the surface as smooth as possible.

Subsequent to the initial compaction, the same workers with the same heated 10-pound tampers began ramming down the asphalt even harder for greater compaction. The rock-asphalt pavement was smoothed a second time with heating irons and tamped a third time. At this point, workers cleaned up the job site and told curious neighbors that they would come back and finish the job in two years.

It was believed that compacting the rock asphalt with a half-ton roller was premature at this point and would shorten the wearing life of the asphalt pavement. Waiting several years for the wear and tear of foot and cart traffic to gradually compress the asphalt, it was incorrectly thought, would give the surface a longer life expectancy. So, years later, a 1000-pound hand roller was brought back to the work site to make

Spreading, raking, and rolling asphalt concrete. Hand tamp is being used on aggregate base in foreground.

Paving was a labor-intensive process in the early 1900s. A Barber Asphalt Paving Company crew paves State Street in Aberdeen, Washington, 1910.

the much-delayed last pass over the surface.

As early as 1863, the manufacturer Aveling & Porter was selling multiton, three-wheel steamrollers. Equipped with steering wheels that looked as though they came from wooden sailing ships, the rollers were reserved for use on earthen and macadamized roadways, likely because of the ill-conceived fear that the prized and costly rock-asphalt roadways of the time weren't up to the punishment of these behemoths.

A PIONEER HOT MIXER

Paving with European rock asphalt didn't make the grade back home in the United States. The high cost of importing the heavy rock that contained precious little asphalt over the Atlantic Ocean was prohibitive. The rock asphalt found in the United States created a street surface that was too slippery. Road builders were left searching for alternatives.

In 1869, while Edward DeSmedt, the Belgian chemist, was putting down his experimental and successful sheet asphalt in Newark, New Jersey, N. B. Abbott was blazing his own trails in Washington, D.C. He studied both mix and method of mixing. In early spring 1871, the first U.S. patent for a bituminous pavement was issued to Abbott. Proclaiming his own ingenuity, he said, "I have invented certain new and useful improvements in Composition Pavements; this invention relates to that class of composition pavement composed mainly of sand, gravel, ashes, broken stone, etc., mixed with tar or other bituminous material to form an elastic and waterproof road or pavement." However, his patented surface was

not a practical one. Made with coal tar and inconsistent in its mix, his pavement failed. Years later, after being paved with a few inches of sheet asphalt, his foot-thick structures were found to provide a good base.

Before Abbott's time, determining which stones were the appropriate size for a pavement mix meant that workers would place fistfuls of stones onto small screens, and then agitate the screens by hand, rejecting the smaller stones and placing the ones of desirable size into the pile to be baked. Abbott simplified this laborious process by introducing a simple rotary drum screen that increased capacity and therefore productivity. It was successful enough that ones of similar design were being used in asphalt plants 50 years later.

Abbott also credited himself with improving the process of heating gravel. Workmen commonly heated stones in cast-iron pots over open fires in an inconsistent and dangerous attempt to get them hot enough to mix with equally hot bitumen being cooked separately. Workers poured the hot coal tar over the hot stones in their trays, stirring the mix in trays or pots resting on bricks or stones. Mixing a batch of pavement material meant four hours of hot, hard work. After a day of hot mixing, only about 200 square yards could be expected. Worse, at the end of their long and underproductive day, the workers "were as thoroughly roasted as the gravel."

Abbott developed a "revolving gravel heater," a cylinder-shaped dryer that placed a small rotating drum inside a larger stationary one. Hot air from a fire-box below circulated between the two. The blistering air heated the stones in the rotating drum with far more consistency than could ever have been achieved by poking at the gravel with handheld instruments. Workers placed cold gravel into one end of a machine and watched hot rocks come out the other. To heat larger amounts of asphalt, Abbott then introduced an iron box with a convex bottom that allowed for a faster, more uniform heating process.

It is not clear whether Abbott assembled all of his inventions in one place, but it is certain that he contributed to what became the future asphalt plant. Unknowingly, Abbott's equipment innovations set the stage for a fellow Washingtonian, Amzi Barber, by helping to make the equipment used by the Barber Asphalt Paving Company more productive. The company that produced many of Abbott's experimental contraptions, Pioneer Iron-Works of Brooklyn, New York, went on to produce many of Barber's asphalt plants and much of his equipment.

In addition, the fact that Abbott's pavements failed was attributed—by the promoters of Trinidad lake asphalt—to the fact that they utilized coal tar. "That

Opposite:

Obispo asphalt, mined from deposits in California, was used to pave eight miles of Canal Street in New Orleans.

Abbott did not understand proportioning is indicated by the fact that his patents did not anticipate the Warren patents some 30 years later; however, had Abbott produced satisfactory surfaces, whether he patented them or not, there would have been nothing for Warren to patent. In any case, had Abbott been successful, tar would have been much more prominent in the United States, Trinidad asphalt would have been much less prominent, sheet asphalt might never have been developed, and there would have been no Warren patents," according to an authoritative text.

It is not known whether Abbott was aware of the comprehensive hot-mix asphalt production facilities that had been developed by Cummer in England and Scotland. Cummer moved his company to Cleveland, Ohio, in 1870 and became the dominant hot-mix plant manufacturing company in the U.S. for many years. Hetherington & Berner entered the business in 1885 and other manufacturers soon followed.

POWER ROLLERS

Power rollers were first known as macadam rollers and then as asphalt rollers and, for a long time, as steamrollers. Although 60 years have passed since the last steamrollers were made and rollers of today are driven by diesel and gas engines, power rollers are still commonly referred to as steamrollers.

Showstoppers on road projects since Roman slaves and soldiers used cylindrical soil/stone compacters, road rollers are still a popular attraction for spectators of construction projects. Rollers were eventually used in the construction and maintenance of macadam roads, and even when asphalt was being laid in the mid-1800s, the new steam-driven power rollers were reserved for macadam surfaces. However, once Abbott and others saw the benefits of compacting asphalt, especially while it was still hot, power rollers became known as asphalt rollers.

A horse-drawn roller produced by the Good Roads Co. had detachable front wheels that could be steered.

Horse-drawn rollers were a tempting concept but an ultimate no-go. The problem was that the horses dug their hoofs in to gain traction, denting the asphalt surface that they were employed to smooth. In 1862, William Barford of Britain was granted a patent for his hollow-steel-drum horse-drawn rollers that he filled with water in hopes of increasing compaction. But water or no water, hoof prints marked the death of horse-drawn rollers.

The very next year, the British manufacturing company

A steamroller, manufactured by Aveling and Porter, on trial in Hyde Park, London, circa 1866

Aveling and Porter demonstrated the first successful steamroller. A large, three-wheeled iron power plant with a tall stack belching black smoke, it demanded attention. It was destined for asphalt, but the benefits and procedures for compacting asphalt were still being explored, and the machine was for the most part left to compressing macadam roads.

Hand rollers of stone and iron were standard equipment for a time to come. These hand rollers had a fire pot in the center to heat the roller. The heat enhanced compaction by keeping the asphalt malleable and helped prevent the asphalt from sticking to the rollers. Quarter-ton, half-ton, and one-ton hand rollers were used successively, so that the compaction became greater with each pass of the next roller. Going from light to heavy helped prevent overstressing the freshly laid pavement. The men pushing and pulling these rollers wore wooden sandals or clogs for insulation and protection.

Two Aveling and Porter steamrollers were shipped to New York City for use in Central Park. The city's commissioners were impressed with their performance and even more stirred by their potential cost savings. On a typical workday, each steamroller was capable of rolling and compressing as much roadway as a horse-drawn seven-ton roller, but the steamrollers didn't need the food and intensive care required by a team of horses. Costing the city of New York just $10 a day to operate versus the $20 a day it took to operate a team of horses, the steamroller enjoyed an indisputable advantage and marked the arrival of the age of steam for road builders.

When it came to asphalt, there were problems with the British steamrollers. The drums on the rollers were too narrow and left marks on the surface of the pavement. Solving this dilemma in 1869 was a three-wheeled steamroller designed in the United States by John Roach and manufactured by Pioneer Iron-Works, which left a seven-foot-wide path of flattened and compressed asphalt

pavement. In 1875, Andrew Lindelhof patented his impressively heavy 10-ton tandem roller, designed especially for asphalt surfaces. But like other steamrollers, Lindelhof's rocked back and forth with the thrusting motion of its pistons, leaving a less-than-smooth surface in its path. Regardless, improvements to steamrollers and their cost advantages were making them a favorite and mainstay of asphalt paving

Improvements to steamrollers made them a favorite and mainstay of asphalt contractors.

contractors by the end of the 1800s.

By 1900, Lindelhof's two-wheeled rollers were commonly called asphalt rollers and were identified with compacting hot-mix asphalt roadways. Three-wheeled rollers like John Roach's were referred to as macadam rollers and were used in the construction and maintenance of hard gravel roads. Asphalt rollers or macadam rollers, they were all steamrollers. Diesel and gas engines were still in their experimental phases, especially in the emerging, barely existent world of road-building equipment.

American road builders were beginning to make an important distinction between themselves and their British counterparts by thoroughly compacting their asphalt surfaces with asphalt rollers, and major advances in equipment improved their efficiency. In 1902, the firm Kelly Springfield manufactured a

Rural roads were the last to be paved, but improving them was essential for farmers to get their goods to market. Spraying liquid asphalt over a stone base controlled dust and helped the road stay dry during rainstorms.

smoother-running and more powerful steamroller that reduced rocking. In 1908, Barford and Perkins built the world's first "solid injection" high-powered diesel engine, marking the beginning of the end for the steamrollers. In 1911, Aveling and Porter outfitted their rollers with a gear that reduced the time it took to change from forward to reverse, thus eliminating problems from asphalt that became cold before it was compacted.

At the start of World War II, the last steamrollers were being manufactured in Britain. About the same time in the United States, Buffalo Springfield Rollers began building rollers with three axles that could be better guided from left to right by a steering action of one of its drums, helping the roller compact the road regardless of its contours. Still, the middle drum was too small and resulted in uneven compaction. Not until three-drum rolling machines were outfitted with larger rollers were they accepted.

EARLY HOT-MIX PLANTS

The first asphalt plants in the U.S. were smelly, loud, dusty places. Found only in large cities, they were located on the edges of business districts and were thus never far from residential neighborhoods. Even though their production levels were low compared to future asphalt plants, the plants' presence was obvious to those nearby. The early asphalt plant's power was provided by a coal-burning steam engine that belched clouds of grit and soot. Hundreds of gallons of oily asphalt boiled in large kettles formed pungent odors that only an asphalt man could love. Drying rocks in large cast-iron drums created a racket that was exacerbated by the ensuing dust storms produced by the shoveling and drying of thousands of tons of stone.

The Warrens didn't just pave roads, they manufactured equipment for paving. One of their early factories is shown.

Looking like casks of wine, barrels of asphalt were delivered to asphalt plants, rolled in on their sides, and left near the heating kettles. Plant workers broke open the barrels and dumped the congealed asphalt onto the ground. They then shoveled the asphalt into the small buckets used for hoisting the asphalt into the heating kettles.

The buckets of asphalt were lifted via ropes and pulleys and dumped into one of two large heating kettles. A boiler man stoked the fires to get the asphalt up to 350 degrees Fahrenheit—it was believed that higher temperatures would burn and damage the asphalt. The wooden staves from the asphalt barrels were burned as fuel for the heating of the kettles.

Rail shipments of sand and small stones, the primary components in the sheet asphalts that Barber Asphalt Company was famous for, were dumped in a heap next to the railroad tracks and as close to the asphalt plant as possible. There they were separated by size for their eventual mixing. A teamster, his team of horses pulling a wooden wagon with a road-grading attachment that looked like a farmer's plow, worked at moving the pile from the edge of the railroad tracks to the mouth of the cold elevator. Joining the action were workers armed with shovels and hoes. Together the crew endured brutal days of dragging, shoveling, and dumping the sand and stones into the constantly rotating elevator.

In 1899, the Barber Asphalt Company proclaimed that it had laid down 11 million square yards of sheet asphalt in the United States as compared with Europe's 3 million square yards of rock asphalt. This was possible, in part, because of the company's automation. The company did not make its own plants or even the components for them, but its famous research and development efforts advanced the asphalt plant's evolution and increased the firm's production of sheet asphalt as well as its mastery in laying it down on the road.

It was a heady time for Amzi Lorenzo Barber, the Asphalt Tycoon. By the standards of the time, his sheet-asphalt plants were highly productive; but an asphalt dynasty was on the rise, and the Warren Brothers Company's plants that produced Bitulithic were becoming even more productive. The large Bitulithic plants put out 60 tons an hour, while sheet-asphalt plants of equal size produced only 36 tons an hour. Semiportable Bitulithic plants were making about 12 tons per hour versus semiportable sheet-asphalt plants that were putting out roughly nine tons per hour. Labor costs were also higher with sheet-asphalt plants, which needed about 16 full-time workers in contrast to about 11 men running a Bitulithic plant.

In the ensuing decades, Barber's sheet asphalts with their sands and very small stones were replaced by the Warren Brothers' two-inch-aggregate–based Bitulithic, which held up better under higher traffic volume and, more important, under heavier vehicles.

In the ensuing decades, Barber's sheet asphalts with their sands and very small stones were replaced by the Warren Brothers' two-inch-aggregate—based Bitulithic.

PLANT NO. 1

Before his death in 1905, the head of the clan, Frederick Warren, noticed that sheet-asphalt pavements were being punished by the increasingly heavy vehicles traveling over them. At first, the wheels of carriages and then later the steel wheels of delivery trucks were cutting ruts into Barber's sheet-asphalt pavements. The simple observation led to the ingenious introduction of Bitulithic. Replacing lake asphalt and sand with refined asphalt and two-inch rocks, Frederick needed one more component for complete success: a factory to make his newly patented hot-mix asphalt.

On April 14, 1903, a patent was issued to the Warrens for what the two-year-old family business proudly called Plant No. 1. On the same day, the Warren Brothers were awarded another patent, a critical follow-up invention for a bin that allowed plant workers to grade and separate the various aggregates used in their large stone mixes. The plant was so advanced and well-built that it became a model for asphalt plants for many years.

On Potter Street in East Cambridge, Massachusetts, across the Charles River from Boston, Plant No. 1 was intentionally located on a site abutting the Cambridge Gas Works facility, which promised an abundant supply of coal tar to be used instead of asphalt for binding the hot mix. The river and a canal leading directly to the plant assured the most efficient and cost-effective delivery of aggregate.

Later, coal tar was replaced by asphalt from petroleum supplied to the plant by special railroad cars. A rail spur allowed delivery of large shipments of asphalt into the plant. Instead of hundreds of wooden barrels of lake asphalt being rolled into the plant and broken open, one railcar could easily and quickly unload up to 10,000 gallons of nearly pure asphalt.

When an asphalt tanker arrived in the yard of the plant, workers hooked steam hoses to nozzles on the tanker. Scorching vapors were injected through the tanker's internal coils—installed for just this purpose—liquefying the asphalt, which had become solid in transport. Once in a liquid state, the asphalt was siphoned out of the tanker and into the plant's own storage tanks.

At first glance, Plant No. 1 was an awkward assemblage of tanks, stacks, drums, conveyors, and furnaces. A closer look revealed poetry in motion. Aggregate received by either barge or rail was loaded into the external fire dryer. After tumbling dry inside a cylindrical drum, the hot rocks were lifted from the dryer to a rotary screen. The rotary screen automatically organized the freshly heated stones by size, dropping them into their designated internal compartments. The newly patented hot-bin separator was capable of holding nine cubic yards of stones.

The still-piping-hot aggregate was sent to the weigh box on an as-needed basis. Earlier weigh boxes that measured stones by volume were inaccurate. The Warrens' weigh box measured the stones by weight and could hold a cubic yard of material. When it was time, the mixing man tipped the box's hot stone contents into the plant's pug mill.

From the other end of the plant came the asphalt. Using an asphalt dipper— a five-foot-long iron pole with a two-gallon bucket affixed to one end—the mixing man poured liquefied asphalt from the melting kettle into the tilting bucket. Attached to a trolley cable system, the tilting bucket was sent to the pug mill, where it was tipped and emptied.

At the center of the plant's operations was the pug mill. Taking about 1000 pounds of stone from the weigh box and 350-degree asphalt from the bituminous tilting bucket, the pug mill looked like a simple container. However, a complex array of internal paddles moved by steam-driven shafts blended the final product—a hot mix—in less than a minute. Beneath the pug mill, if everything was going to plan, was a wooden wagon ready to receive the half-ton load of hot mix.

During the first 20 years of the 1900s, the largest sheet-asphalt plants, which required nearly two minutes for their pug mills to mix a batch, were producing about 36 tons an hour. The Warrens, with a shorter mixing time and larger aggregate, were reporting nearly 60 tons per hour—a 67 percent higher production rate than Barber Asphalt Paving Company and others.

By 1907, the Warren Brothers had manufactured 40 permanent plants for themselves and other asphalt contractors. Loyal to the end, Plant No. 1 chugged along until the Warren Company, at one time the largest asphalt producer in the United States, was sold in 1966. The stationary Plant No. 1 was their first, but the Warren Brothers were soon building semiportable and portable plants. Feeding a nation hungry for paved roads, the company's portable plants helped deliver asphalt to where the demand was greatest—rural America.

SEMIPORTABLE PLANTS

In the early years of our country, every state capital was located on a waterway— harbor, bay, or river. In the 1800s, access to waterways made it possible to ship the first heavy rock asphalts from Europe and later the lake asphalts from Trinidad to these primary cities. Smaller cities off the main shipping channels or towns only accessible by railroads or dirt roads were out of luck. Even if asphalt could be shipped to these remote locations, only the densely populated and prosperous cities could rationalize the expense of building permanent plants. Getting asphalt to pave the streets of the small towns had to wait for semiportable and portable plants.

A 1908 rail car–mounted asphalt plant built by Hetherington & Berner was powered by a steam engine mounted at one end with a belt drive similar to the threshing machines of the period.

Taking all the parts of a larger plant and simply mounting them onto railroad cars, the Barber Asphalt Paving Company claimed that it had found "the solution to the paving problem in smaller cities." The range of hauling hot asphalt from a plant was severely limited to three miles, so it was necessary to bring the plant to the small town. Semiportable plants were itinerant factories on wheels.

The challenge in the late 1800s and early 1900s was moving the semi-portable plants. Only after much assembling and disassembling of many cumbersome components mounted on two large flatbed railroad cars could a small army of workers relocate an asphalt plant. Sometimes, an asphalt contractor would even lay a rail spur a mile or two to a paving site.

In 1899, the Barber Asphalt Paving Company was operating hot-mix plants requiring two 60-foot flatbed railcars. One car was equipped for handling aggregate, with a conveyor to lift sand and small stones, a rotating screen to filter them, a drum dryer for heating them, and a hot elevator and hot bin to hold them before their mixing in the pug mill. The first car also had a large boiler for creating steam and heating the asphalt kettles located on the second car, which was less cluttered but equally important. On it sat large asphalt-melting kettles and the plant's main source of power—a large coal-burning steam engine that provided the energy needed to drive the conglomeration of shafts and belts.

In a space created between the two cars, the pug mill was erected on wooden trusses with just enough room below it to park a horse-drawn wagon, or later a truck, and load it with hot mix. A wooden platform erected between the two cars housed the measuring box for stone and sand and the tilting batch bucket for pouring the molten asphalt from the melting kettles. The pug mill mixed the sand with the asphalt and minerals before fluxes for the sheet asphalt were added into the pug. The contents of the pug mill were then dumped into a waiting wagon.

The plants were not pretty, but they were productive, putting out 12 tons an hour. Most important, they were portable.

By 1920, the Warren Company proclaimed that it had manufactured almost 80 semiportable plants, more than twice as many as all of their competitors combined.

THE TRULY PORTABLE PLANT

Increasing portability twofold, semiportable plants were eventually mounted onto one railcar instead of two. Still, towns off the rail lines were left without access to asphalt plants, and as a result they were stuck in the mud until the Warren Brothers patented yet another revolutionary product—a truly portable asphalt plant. That patent, issued on July 8, 1913, was for a portable plant mounted on a trailer that could travel over the open road. It set off two revolutionary movements. First, plants could now be pulled, albeit with enormous

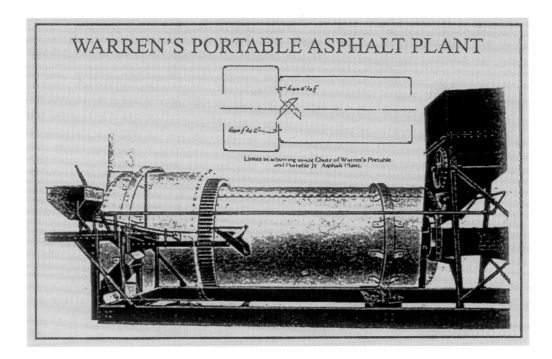

WARREN'S PORTABLE ASPHALT PLANT

Limits in adjusting inside Chute of Warren's Portable and Portable Jr. Asphalt Plant.

Towns off the rail lines were left without access to asphalt plants . . . until the Warren Brothers patented yet another revolutionary product—a truly portable asphalt plant.

effort, to remote locations not serviced by a rail line or waterway. Second, the plants' relative affordability and portability encouraged smaller start-up companies to enter the asphalt-paving business.

As work progressed along a road, the Warren Brothers' portable plant was pulled along behind the road gangs by a steamroller or by the new "caterpillar tractor" that had first been successfully tested in 1904 by Benjamin Holt, one of the founders of Caterpillar Inc. Introduced to the market a couple of years later, the newfangled steam-driven tractor was designed for farming but was quickly put to work by road-building contractors.

Invented by H. W. Ash, the portable plants were called Warrenite plants after a newly patented mix called Warrenite that, like the plant itself, was designed for paving rural highways. Roughly a dozen men operating the plant could produce 15 tons of hot mix an hour.

A dynamic feature of the Warrenite plant was its internal fire dryer. Nearly five feet in diameter and nine feet long, it was fired by an oil burner that became a model for other dryers. The fire was contained inside the dryer, reducing drying time and consuming less fuel. External fire dryers, whose fires needed to penetrate the dryer's thick iron sides before the stones could be heated, were no match. It had been incorrectly believed that flames coming into direct contact with the aggregate would damage the stones. The Warren Brothers spent the next decade overcoming that myth before their ingenious design was accepted.

The little plant had other state-of-the-art features, like a scale featuring a dial instead of a long, clumsy bar and counterweight. The melting kettle used steam pressure to force asphalt out of the kettle and into the tilting bucket for mixing.

RETROFITTING MIXERS

In 1910, asphalt contractors took concrete transit mixers and began adapting them to work as portable pug mills. To heat the mix, the contractor built a hood around the entire drum mixer, so that hot gases from a firebox beneath it were trapped and contained.

Going a step further and displaying chutzpah, if not recklessness, contractors began placing kerosene torches in the mixer. Cranking up the heat even more meant that one cubic yard could be mixed at a time. These souped-up rigs produced a usable hot mix at a respectable three tons an hour.

Concrete mixers were adapted to asphalt, producing up to three tons of hot mix an hour.

Drying the stones before they went into the retrofitted concrete mixer was the contractors' biggest challenge. Meeting that challenge a couple of years later,

the Warren Brothers produced a dual-chamber drum mixing plant. The first of the two compartments heated and dried the aggregate. The second area acted as a pug mill, blending the heated stone with the heated asphalt. Hot gases traveled through both drums, giving enough heat to dry the stones and create a hot mix. Part drum dryer and part mixer, this plant attempted to miniaturize the primary components of an asphalt plant so that a small operator could afford to pave roads in remote areas.

An entire plant was needed to perform the task of making a consistently usable hot mix, but the conversion of concrete mixers into hot-mix asphalt machines further proved the willingness of early asphalt contractors to invent, reinvent, and improve themselves.

THE GOB BOX

Gob boxes were one of the simplest of all the moving parts of the plant. Looking like a wooden bottom-dump wagon that had been unhitched from its team of horses and affixed beneath the pug mill, the gob box simply held a "gob" of asphalt equal to the capacity of the pug mill. Once the pug mill completed its minute or two of mixing, its contents of hot mix were deposited in the gob box, which held the mix until a team of horses, or in rare cases a truck, was in position beneath it. The first gob boxes had trapdoor mechanisms similar to those of the bottom-dump wagons positioned beneath them. With a pull of a lever, the gob box's two doors opened and its 1000 pounds of asphalt poured into the vehicle below. Providing short-term storage for just-produced asphalt, the gob box allowed the pug mill to immediately take on an additional load of asphalt and aggregate from the production line. It kept production one batch ahead of the trucks.

After World War I, U.S. Army surplus trucks returning from overseas were converted into dump trucks and Model-Ts retrofitted with dumping containers took up their positions under the gob box.

THE HAUL

Long in time but short in distance, the haul from the plant to the job site was at first a hard and time-consuming trip. As wooden wagons were replaced by retrofitted Model-Ts and dump trucks, hauling asphalt made its way from antiquated to automated by 1920.

Up until this time, a haul consisted of a horse pulling a wooden wagon with a load of asphalt from the plant to the crew laying it down on the road. The bottom-dump wagon released the asphalt onto the street to be paved. The dilemma of slow carts was exacerbated by their small loads—the smaller the pile of hot mix, the faster it cooled, so the early wooden wagons were limited to a haul of three miles or less.

By mistake, it was learned how effective larger loads of hot mixes were at insulating themselves. In 1899, an industrial experiment was carried out unintentionally when 300 tons of piping-hot asphalt were loaded in Long Island City into a flat-bottomed scow commonly used for transporting ore, aggregate, and sand. The hot asphalt was hauled to a work site in New Rochelle, nearly 20 miles away; but when the barge arrived, so too had foul weather. The crew was forced to pull large tarps over the hot mix and wait for the rain to stop.

Thirty hours later, it was expected that the 300 tons of asphalt would be too cool to lay down and therefore a complete loss. To everyone's surprise, the disaster turned into an industrial breakthrough. Except for a hard crust on the top, the entire haul was applied to the job. The incident became the first known example of how hot mixes could be stored after production.

Trucks that were capable of moving faster and carrying larger loads than horse-drawn wagons began making their mark by 1920. Expanding the range of delivering asphalt by more than twofold, trucks could work all day without a meal, and they offered a smoother ride because they were built with better suspension systems than their wooden wagon ancestors.

The motortruck delivered more asphalt that was hotter and none the worse for wear over the road. Despite the truck's inevitable dominance, mules and horses were seen on job sites up to the end of World War II.

THE LAYDOWN

Arriving at the job site, teamsters were directed by a man known as the wagon dumper. Positioning the bottom-dump wagon over a dump board—a sheet of metal or wood laid flat on the unpaved roadway—the wagon dumpers saw to it that the load in the wagon landed squarely on the dump board. The teamster then headed off to the asphalt plant for his next load.

Stabbing the steaming heap of asphalt with their shovels, a team of five men dispersed the pile over the freshly swept cobblestone or old macadam roadway. The dump boards and their loads of asphalt were attached to mules by sets of

Dumping the hot-mix asphalt was just the first step in paving. Shovelers and rakers spread the material evenly before the steamroller compacted the new road.

chains. Dragging the dump boards, the mules led a parade of 15 men and at least one steamroller. They were a ragtag group wearing wool-brimmed hats to keep the sun from their eyes and walking over vaporous, ink-black material with shoes of wood while wielding odd utensils.

The mule may have been leading the gang, but the foreman was the indisputable boss. Having acquired the skills of each of the men under his watch through years of toil, the foreman would pitch in, shoveling, raking, and rolling alongside his men wherever and whenever needed.

On the heels of the five shovelers were two rakers. These men spread out the hot mix as evenly as possible, giving the road a well-graded, three-inch-thick coat of asphalt. Moving as quickly as possible, the rakers made sure that lumps in the hot mix didn't consolidate into what would otherwise become permanent bumps in the road. The rake was an exception to a rule of no wooden tools. At the end of its steel shank was a wooden grip. The combination of steel and wood meant less weight and more control for the craftsman wielding the instrument. The only other exception to the no-wood rule was the sweeper's push broom.

Finishing the job, the man on the left is operating a heated smoothing iron to take out any bumps left by the paving process. The workers on the right are spreading sand on the surface.

Omnipresent, the sweeper rid the road of debris.

Wooden tools could burn under contact with fires from hot mixes, furnaces, and melting kettles. This was especially true for workers laying sheet asphalt, as these crews were serviced by fire wagons—smoldering contraptions that kept the steel tools hot all day long to avoid having the bituminous materials congeal and harden on the instruments. Whenever a worker's tool became gunked up or when he left for a break, the iron rakes, shovels, or tamps were placed in the fire wagon. Unfortunately for the workers, some of the iron rakes were not only hot but heavy, with some weighing nearly 70 pounds.

A laydown job might be ruined if the workers walked on the freshly laid hot mix without specially made wooden clogs or sandals fastened to their boots with leather straps. Not only did the fire-retardant footwear prevent hot feet, it allowed the men to walk on freshly laid asphalt without damaging it.

After the hot mix was dumped, shoveled, and raked, it was tamped with an eight-inch-by-six-inch plate attached to a five-foot-long iron handle used to smash down rises, bumps, and lumps. If necessary, a laborer with a smoothing iron heated spots of asphalt that had cooled, so that they too could be flattened appropriately.

The roller men were the captains and commanders of the most-watched pieces of equipment in town—the steamrollers. Working quickly so as not to let the asphalt cool too much, the first, lighter rollers—weighing two and a half tons—conducted "break-down" rolling while the asphalt was still about 280 degrees Fahrenheit. Before steamrollers were common, break-down rolling was performed by hand rollers weighing 500 to 1000 pounds. About two feet wide, each was equipped with a seven-foot-long handle with a T-grip at its end. Donning wooden footgear, a roller man would walk over the hot mix, pulling or pushing his instrument by hand.

After allowing the asphalt to cool a bit further, a heavier roller passed over the pavement. The second pass was made with a steamroller weighing between five and seven tons. This was timed for the period when the temperature of the mix was falling below 200 degrees Fahrenheit. In a cooler state, the asphalt was more stable and wasn't prone to lateral displacement, the horizontal shifting of pavement over the ground.

Hot mix sticking to the giant steel wheels, called pickup, was more than just a nuisance. Pickup could cause divots in the road's new surface, requiring the crew to rework it. To avoid this, the drums were kept wet with rags drenched in water and draped over the wheels. Later, scraper blades were attached to the surfaces of the wheels. With each rotation, the scrapers kept the roller free of pickup and the roadway smoother. A dusting of limestone or Portland cement powder might be spread over the surface of the new pavement, to reduce pickup to a minimum.

Around 1900, cross-rolling became popular with contractors. With one roller running up and down a street in the direction that traffic would move, a second roller would run from side to side across the street or at a 45-degree angle to the curbs. Cross-rolling was believed to offer the asphalt a longer life.

THE ROAD AHEAD

Into the 1920s, hot asphalt mix was hauled by teamster and team. Once it arrived at the site it was dumped, spread, and smoothed by hand.

A gap between the speed at which a hot mix could be produced and the speed it could be hauled, laid, and rolled plagued asphalt producers and paving contractors. The processes of hauling and laying down pavement were lagging behind the development of asphalt plants, slowing the progress of getting America out of the mud.

Not until the late 1920s and 1930s, when trucks quickened the hauling process and new inventions for laying down hot mix were introduced, was the gap closed.

When America saw the advantage of shipping massive amounts of goods via trucks, the roads suffered under the heavy loads.

CHAPTER EIGHT

Attacking Asphalt with Tank and Truck

"Unclassable, almost impassable, scarcely jackassable!"

First East Bay Ship By Truck Tour.
May. 17 - 21 — 1920.
Robert W. Maytland — Train Comman

MAKING LIGHT OF BAD SITUATIONS, activists at the turn of the 20th century sang and joked about the conditions found on the nation's highways. Later, however, when Americans went to war in 1917, passing over rural highways wasn't a luxury but a necessity, and the jokes stopped. It was difficult to determine which suffered most—the roads, the trucks, or the truck drivers.

At the beginning of the 20th century, Americans began shipping goods by truck, taking deliveries by truck, and building highways with trucks. By the 1920s, trucking had come to stay, creating a new industry that improved the quality of life for Americans by speeding up deliveries, reducing shipping costs, and creating a competitive alternative to the railroad industry. The truck, alongside the tank, was a new weapon in the nation's armory. Trucks and the emerging trucking industry were another reason roads needed to be built. Trucks became a central issue in the campaign to strengthen the nation's network of roads.

Since the 13th century, trucking has meant the act of swapping, giving, bartering, exchanging, or trafficking in goods. The truck as an object was anything from a single-wheeled barrow or a two-wheeled pushcart to a small-wheeled gun carriage.

These steam-powered trucks were among the earliest produced by the White Motor Company, in 1900. The company switched to gasoline engines in 1910.

on up to the groups of wheels underneath a railroad car. Starting in the early 1900s, a new form of truck, the motortruck, forever changed the way Americans lived and conducted business.

More than any other type of road-building equipment at the time, the motorized gasoline or diesel truck shifted the methods of road builders. In the world of asphalt suppliers and paving contractors, trucks increased production, permitting more hot mix to be shipped in less time. More tons of hot mix on the roads meant that the motoring public—in both trucks and cars—could move over them at faster speeds. In the 1920s, road building was still a horse-and-wagon venture; but by the 1930s, the industry was using dump trucks, flatbed trucks, and tractor trucks.

In the early 1900s, however, the truck was loathed as much as the automobile was loved. Automobiles were about status and adventure. Trucks were about hard work and a hard living. It was a thankless beginning for the workhorse vehicles, which were mostly relegated to the short hauls that the railroads couldn't be bothered with. In the cities, trucks mostly performed local deliveries of household items—furniture, fruit, and heating fuels. On the farm, trucks hauled goods between barn and field or from farm to town. Slow and underpowered, trucks blocked the flow of pedestrians, horses, and automobiles, fighting it out for space on narrow and congested streets and roads.

Soon, however, the public turned to trucks. As time went on, they began to outperform the horse and wagon and were accepted for use in matters of life and death. Medical doctors were among the first to employ motorized vehicles as professional tools, choosing them when minutes could save a patient's life. Fire

In the early 1900s, the truck was loathed as much as the automobile was loved.

departments followed, purchasing trucks to replace horse-drawn fire-hose vehicles and ambulances. Later, trucks helped turn the tide of World War I from near defeat to victory.

The San Francisco earthquake of 1906 was the motor truck's world premiere as emergency vehicle on a grand scale. With the city in ruins and on fire, San Francisco government officials commandeered 200 private trucks and automobiles for emergency purposes. Devouring 15,000 gallons of fuel donated by the Standard Oil Company, the trucks and other vehicles performed heroically while horses died of exhaustion. In the end, the truck was credited with saving the city from complete destruction. City officials testified that trucks had moved the sick and injured to shelter and helped keep law and order nonstop for three days.

In 1911, interstate trucking began in the United States with the completion of the first transcontinental truck crossing. In March of that year, a truck called the Pioneer Freighter, built by the Saurer Motor Car Company of New York and weighing three tons, propelled itself across America with a tiny 37-horsepower gasoline engine. Looking like a horseless covered wagon, the truck crawled its way through the southwestern United States between Colorado and California on stiff solid-rubber tires. The Pioneer Freighter carried three and a half tons of

In 1911, a truck called the Pioneer Freighter, built by the Saurer Motor Car Company of New York and weighing three tons, propelled itself across America.

lumber so that the expedition's crew could build their own bridges over creeks, swamps and railroad tracks. After traveling 1450 miles, halfway into their trip, the crew could only claim the average speed of a slow horse: 3.26 miles per hour. The sad state of the country's unpaved roads kept progress to a crawl.

The Pioneer Freighter's expedition was led by A. L. Westgard of the American Touring Club of America. Westgard, whose members nicknamed him Pathfinder, was commissioned as a special agent of the Office of Public Roads, the federal government's official highway division. Although he received no pay for the honor of exploring the feasibility of interstate trucking, Westgard led his expedition out of Denver in March, traversing the desert into Santa Fe, New Mexico. From there, the crew followed the basic path of the yet-to-be-named U.S. Route 66 going into the White Mountains of Arizona through Phoenix, finally arriving in Los Angeles. Once on the Pacific Coast, they loaded the Pioneer Freighter onto a train back to Colorado, then pointed it toward the Atlantic Ocean. The Pioneer Freighter truck headed east under its own power and arrived in New York City in July.

The following year, a farsighted U.S. Army captain, Alexander E. Williams, convinced his commanding officers that it was time to replace the calvary's horses and mules with what some were calling modern warhorses—motortrucks. The 1912 Army Road Test, from Washington, D.C., by way of Atlanta, Georgia, to Fort Benjamin Harrison, Indiana, was the 1503-mile endurance test. Williams was sent off with the words of an old-school general who told him, "When your trucks get stuck, just wire me and I'll send you some mules!"

The roads were primitive, the winter weather was harsh, and to complete the nightmare, the trucks were fully loaded with supplies and a half ton of sand for good measure. Entries into Captain Williams's log of observations while en route included, "Road so bad a horse drowned in mud hole before it could be unhitched"; "Red mud axle deep. Had to leave road and travel cross-country trail"; "Built bridge to cross stream"; "So muddy, even got stuck on downgrades"; "Used telephone poles and block-and-tackle to get trucks through mud." Three of the four trucks made the entire route, and the daring

This one-and-a-half- ton truck, manufactured by FWD in 1912, was originally used by the U.S. Army before it became a commercial vehicle.

captain had convinced the army that trucks couldn't be ignored.

By 1914, The United States had nearly 2.5 million miles of roads for trucks to roll along, but only 14,400 miles were well paved—10,500 with an asphalt surface, 2300 miles built with concrete, and 1600 miles built with red bricks. In other words, 99.5 percent of the nation's roads were crude dirt roadways.

TRUCKING'S SACRED PATH

Events in Europe were also changing the way the world viewed motortrucks and the highways they traveled over. World War I was raging in France in 1914, and American truck manufacturers were getting orders from the Allies. In the Battle of Verdun, France, one of the deadliest of the war, the French sent 8000 trucks into action, carrying reinforcements, ammunition, and supplies to the battlefront and ferrying injured soldiers back. During the crucial battle, trucks were dispatched at an astonishing rate of one every 25 seconds. The highway they used during the campaign was called the Sacred Path as a testament to its

In the Battle of Verdun, France, one of the deadliest of the war, the French sent 8000 trucks into action.

General Motors eventually produced 50 of these ambulances a day. They were the army's standard version in 1916–17.

significance. Today it still carries the name. Thousands of trucks and the Sacred Path highway enabled the French to resist one of the most intense German invasions of the war.

The United States declared war on Germany on April 6, 1917. American industries prepared for the conflict as President Woodrow Wilson called for immediate and complete mobilization. Within the year, the President had taken control of the national railroad system to bring order to the flow of goods for the war effort. Supplies and munitions headed for Europe took precedence over all other rail cargo, displacing basic necessities like milk and fresh produce that were summarily removed from shipping schedules, leaving farmers without a way to get food into the cities. The country had become dependent on a single mode of transportation: the railroads. Yankee ingenuity and innovation arrived in the form of the truck, eventually resolving a national crisis.

It became patriotic to "Ship by Truck," as one national campaign urged. Farmers, desperate to get their crops to market before they rotted in their fields, loaded up their farm vehicles and drove them into the city to reach their markets. Since many farmers did not have their own trucks, merchants sent their fleets' drivers out of the cities and into the country for the first time. Almost instantly, interstate trucking became an industry as routes opened between Philadelphia and New York and between Toledo and Detroit.

Trucks not only succeeded in picking up the overflow from the railroads, they also streamlined the overall shipping process. Eliminating the railroads from the equation meant that merchandise could travel directly from farm to market, reducing the likelihood of breakage during shipment and spoilage while stored at railroad warehouses. Trucks performed better on the short hauls than trains, resulting in a loss of business that the railroads were not destined to recover.

Trucks had heavy bodies and small engines, so they were susceptible to getting stuck in the mud and were unable to climb steep grades. Autos, however, had relatively light bodies and strong engines, allowing them to get through mudholes and up hills. Automobiles had another major advantage: they often had pneumatic tires. Providing cushions of air and giving cars better traction than the steel or solid-rubber tires used on trucks at the time, pneumatic tires afforded passengers a softer ride at higher speeds. A strong air-filled tire capable of supporting a heavy truck was not available until Goodyear Tire Company brought one to market in 1916. Steel and solid-rubber tires limited trucks to top speeds of nine miles per hour. They also pulverized the roads they traversed.

On April 9, 1917, three days after the United States entered the fighting in Europe, Goodyear launched the Wingfoot Express, a nonstop service between Akron, Ohio, and Killingly, Connecticut. The first truck rolled out of Akron with a load of new tires and headed for Connecticut. Twenty-eight days and 28 tires later, the first run was completed. The return trip was accomplished in seven days because of better weather and drier roads.

Encouraged by their successes and with the potential of enormous profits from the sale of tires if interstate trucking took hold, Goodyear launched the county's first transcontinental truck route between Boston and San Francisco in September of 1918. The patriotic call to Ship by Truck was answered by the inaugural run, which departed from Boston with a pilot car guiding two Wingfoot

A strong air-filled tire capable of supporting a heavy truck was not available until Goodyear Tire Company brought one to market in 1916.

The Wingfoot Express made cross-country runs by 1918. If a wooden bridge collapsed, repairs were made on the spot.

U.S. ARMY OVERSEAS DRIVEAWAY; WOODWARD AVE. 1916

The U.S. Army changed dramatically with the addition of motorized trucks in 1916. The ones shown were sent to the Mexican border, where General "Black Jack" Pershing was pursuing Pancho Villa for his attacks on Americans. Pershing became an advocate of better highways as a result of his experience with motor vehicles there and in World War I.

Express trucks loaded with cotton-cord fabric for tire fabrication in Akron. Arriving in Akron, the Wingfoot's cargo was exchanged for new airplane tires that were eventually delivered to San Francisco. The 3700-mile trip, 70 percent of it over mud roads, took just 14 days, even though the trucks fell through several rotten wooden bridges.

Goodyear's new pneumatic tires increased the speed at which trucks could travel. More important to the tax-paying citizens, the new tires spread the weight of the truck's load over the highway, causing less damage to its surface than the solid-rubber and solid-steel tires. In order to keep the trucks moving through the night, the Wingfoot Express trucks were designed with the first-ever sleeping cabs, a cramped space behind the operator's head. The Wingfoot Express trucks became a rolling billboard for American interstate trucking and a call for better highways.

MOBILIZATION—NOT BOMBS—DESTROYS U.S. ROADWAYS

The United States Army placed an order for 30,000 trucks in September of 1917. That number would not be enough, according to General John J. "Black Jack" Pershing, commander of the U.S. Army in Europe, who estimated that his forces would need at least 50,000 trucks and an endless supply of spare parts in order to defeat Germany.

At the height of the war, General Pershing wrote a critical communiqué to his chief of staff, stressing, "At the present time our ability to supply and maneuver our forces depends largely on motor transportation. . . . The need for motor transportation is urgent. It is not understood why greater advantage has not been taken of deck space to ship motor trucks. Trucks do not overburden dock accommodations or require railroad transportation. Can you not impress this upon shipping authorities?"

Back home, the problem was getting trucks from the Midwest to the ports of embarkation on the East Coast. In times of peace, the railroads would have been the obvious choice; but with the war's demands on them, they were unable to handle a delivery of this magnitude. The army's quartermaster general ordered the 30,000 trucks to be driven under their own power from factory to ship. The challenge: there was not a single marked roadway that could be taken. No one was too sure of the best set of roads or even of a viable route between the factory in Detroit and the port of Philadelphia. There were no numbered routes to follow.

On November 22, 1917, a reconnaissance mission was dispatched from Toledo, Ohio. Its objective was to find a series of dirt roads between Detroit's factories and Philadelphia's shipyards. Disastrous road conditions and the overloaded railroad system were a severe hindrance to America's ability to gear up for World War I. It was now clear that our national defense depended on a strong highway system.

Three weeks after the primary route was selected, the warbound trucks took to the road carrying banners reading, "Detroit to Berlin." Running southeast through Ohio to Pittsburgh, the path ascended into the Allegheny and Appalachian mountains of Pennsylvania before reaching Philadelphia. Secondary routes were used along the old National Road. The trucks then rolled either to the port of Baltimore or to New York City. In one of the worst winters ever recorded in the United States, convoys of trucks began rolling down the main streets of small farming towns—a parade of military might that lasted the

Three weeks after the primary route was selected, the warbound trucks took to the road carrying banners reading, "Detroit to Berlin."

duration of the war. The winter storms blew six-foot-high snowdrifts across the mountain passes, requiring the 24-hour vigilance of highway crews to keep the trucks moving.

Delivering the trucks under their own power freed 17,250 railroad boxcars for the shipment of other war goods. Each truck, loaded with as much as five tons of ammunition and spare parts, was only capable of traveling at 14 miles per hour. Their solid-rubber tires tore through the road surface they moved over. The drivers needed brawn and brains to keep from sliding off the roads. An entire highway in Delaware was ruined by the passing of a single overloaded truck with 11 tons of cargo.

General Pershing kept demanding more trucks right up until Germany's surrender on November 11, 1918. Victory in Europe came, but at the cost of the nation's road network at home. This was not a case of routine wear and tear; it was, in the words of one federal official, the "simultaneous destruction of the entire road system." Almost overnight, the thinly surfaced, narrow roads were wiped out by the steel and solid-rubber tires of the primitive trucks. Every type of surface—brick, stone, gravel, concrete, and asphalt—failed under the trucks' brutal tires and extreme weights.

Despite what the Romans had taught the world about the importance of roads, the modern highway had not been considered as a serious alternative to the railroads and ships. It was World War I that revealed the importance of roads to national defense.

Engineering News Record magazine in April 1918 encouraged the public to address the dilemma, saying, "We are faced at the outset, then, with the question whether it will pay to build better and costlier roads than any yet contemplated."

MOTOR TRANSPORT CONVOY — JULY 7th 1919

© Weller Studio ~ Baltre. Md. Park Bank Bldg

A massive convoy crossing America in 1919 proved the necessity of good roads for national defense.

After the war, having been given undeniable evidence of the roads' worth, Americans picked up the pieces and chose to rebuild the road system. The nation was entering its Golden Age of Road Building.

"THROUGH DARKEST AMERICA WITH TANK AND TRUCK"

"I had missed the boat in the war we had been told would end all wars," complained a resentful young tank corps commander, Dwight David "Ike" Eisenhower.

In 1918, training enlisted men on a Civil War battlefield and with small naval ship cannons for the fighting in Europe was not Ike's idea of how he wanted to spend the critical years of World War I. Without real tanks, Eisenhower was reduced to simulating war situations by mounting machine guns on trucks and storming an imaginary enemy while firing live rounds into the famous ridge Big Round Top at the Gettysburg battlefield.

Eight months after the war ended, Ike volunteered to join the U.S. Army's first motorized military convoy to cross the country. The military planned the convoy as a way to campaign for better highways that it believed were needed for national defense.

"In those days we were not sure it could be accomplished at all. Nothing of the sort had ever been attempted," Ike recalled in his memoirs. Calling the journey "Through Darkest America with Tank and Truck," Ike dedicated an entire chapter of his autobiography to the subject. The 3000-mile trip made its mark on the 28-year-old brevet lieutenant colonel, forever shaping his understanding of roads and their role in a country's defenses. The excursion opened Ike's mind to

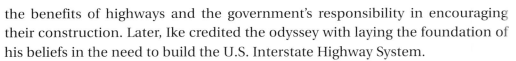

Dirt roads made up more than half of the convoy's route. Trouble was, they often turned to mud, and getting stuck in the mud was a daily occurrence. Below: A permanent Zero Milestone now marks the spot where the convoy started in Washington, D.C.

the benefits of highways and the government's responsibility in encouraging their construction. Later, Ike credited the odyssey with laying the foundation of his beliefs in the need to build the U.S. Interstate Highway System.

On July 7, 1919, Secretary of War Newton Baker ceremoniously proclaimed, "This is the beginning of a new era" and ordered the convoy to "Proceed by way of the Lincoln Highway to San Francisco without delay." This, however, was much easier said than done. One member of the expedition admitted that the Lincoln Highway at the time "existed largely in the imagination and on paper." The convoy pulled out of Washington, heading north for the famed highway, at an average speed of six miles per hour. They departed from the site of the Zero Milestone that today stands on the Ellipse south of the White House. It was the point from which all highway miles in the United States were to be measured. Along the way, at their stop in Frederick, Maryland, they picked up Eisenhower and another officer.

The journey was a struggle from the start. Covering just 58 miles per day on average, it was not the vehicles but the roads that limited their progress.

When under way, the convoy stretched out over three miles of road. An impressive show of force, it included 39 officers and 258 enlisted men driving 81 vehicles that ranged in size and shape from 37 giant cargo trucks weighing up to 10 tons, to 10 midsize delivery trucks, five ambulances, a searchlight vehicle, 11 automobiles for the officers (including a Cadillac for the expeditionary commander), nine motorcycles, four kitchen trailers, and even a four-ton trailer carrying a pontoon boat called the *Mayflower II*.

Both the Lincoln Highway and the convoy suffered. Good luck prevailed and no one was killed, but there were injuries. Trucks crashed through bridges and into rivers. They skidded off roads and rolled down mountainsides, succumbing to the beating by the road. In the end, the convoy surrendered nine trucks to the highway, leaving them behind as unsalvageable. In his log, the ordnance officer on the convoy reported 230 calamities, one for every 14 miles traveled. Ike called that report "the catalog of disasters." The convoy damaged and destroyed 88 bridges by falling through them. The three-mile-long train of machines also tore up countless miles of the Lincoln Highway with its solid-steel wheels.

Toward the end of the trip and trying to make up for lost time, the engineer corps was ordered to move out 24 hours ahead of the convoy to scout the roads, looking for hazards and repairing weak bridges and dangerous sections of road. Ike said, "[T]his arrangement avoided many unnecessary delays."

Two months later, Lieutenant Colonel Eisenhower submitted his official report, saying, "At best, the Lincoln Highway over this portion of the country is so poor as to warrant a thorough investigation of possible routes for building a road, before any government money should be expended on such a project."

The report added, "The truck train was well received at all points along the route. It seemed that there was a great deal of sentiment for improving highways, and, from the standpoint of promoting this sentiment, the trip was an undoubted success." He noted that nearly every officer on the convoy filed reports recommending building public support to construct better roads.

Eisenhower (center of picture in light breeches) learned the importance of an interstate road system when he traveled with the convoy.

HAIL TO THE CHIEF

Acting more like the country's commander in chief than the head of its Bureau of Public Roads (BPR), Thomas H. MacDonald declared war on bad roads. MacDonald's position was "chief of bureau." Under a variety of titles, he would hold the position until 1953, but he was always known as "the Chief."

Traffic regulation was nonexistent on early roads. Note the small car pulling out to pass a large truck.

MacDonald's was a moral war that the federal government and the states would fight together. The Chief proclaimed, "Those who do not have qualities of manliness, square dealing, good temper, and ability to get along with people must go," adding, "This is an All American Job! We must recognize the important part that the states have in the work." This work, he declared, was a "Gigantic Business."

After the war in Europe and the Chief's appointment to the Bureau of Public Roads, the federal government's road-building agency launched a construction campaign. During his first year in office, the Chief and his engineers approved more work than in the previous three years combined. Releasing enormous sums of federal money for the work, his new staff at the Bureau of Pubic Roads approved 90 percent of state highway requests for funding within four days of their arrival at the BPR offices. When thousands of railcars were needed to haul aggregate to prevent a slowdown in highway construction, the Chief successfully negotiated with the powerful Interstate Commerce Commission to make it happen.

The Chief didn't stop there. Securing over 40 million pounds of explosives at a cost of $10 million, he cleared the way to blast away obstacles for new roads.

The states used this stockpile with enthusiasm. As one highway engineer explained, "The results of TNT in rock blasting are so far superior to those of any other explosive that we have found that an experienced powderman who has used TNT can hardly be induced to use anything else." Steam shovels, hand shovels, automobiles and their spare parts, and just about anything else the U.S. Army was willing to part with were used to build roads. The Chief was able to facilitate the use of nearly a quarter of a billion dollars of decommissioned military hardware in the construction of his new interstate highways.

In 1922, at the request of Chief MacDonald, the U.S. Army, under the command of General Pershing, prepared the so-called "Pershing Map," which defined a system of national routes considered important for national defense. Pershing's experiences with the use of motorized battle equipment in the campaign against Pancho Villa and the unbelievable logistical problems in getting trucks to overseas shipping points during World War I drove him to be an avid supporter of improved roads.

Pershing testified before Congress that the most important highways to national defense were also the highways that were most needed for commercial and industrial growth and the motoring public—in other words, the very highways the Chief was seeking support in building. The Pershing Map

The Pershing Map, which in 1922 defined the routes considered important for national defense, became the blueprint for the Interstate System decades later.

of national defense roads, approved on August 23, 1922, heralded the path of what would later become the U.S. Interstate Highway System.

The Chief was not a gifted orator, but everyone in the room listened when he spoke about the greatness of the nation's road-building efforts. He boasted that not since Rome under Julius Caesar or France under the Emperor Napoleon had so many miles of road been built as in the period under his leadership. He proudly pointed out the difference—the first two were accomplished by despots, the latter by a free people operating in a democracy.

CHAPTER NINE

Lying Lightly on the Land

"We pioneered a new road up to Glacier Point in Yosemite. The old road was too rugged and too dangerous for automobiles. It was more of a trail than a road. There wasn't any pavement and the gradients were severe. The National Park people said it had to go."

"The public would follow any road into the wilderness, so long as it was paved," was the belief of Stephen Mather, the first director of the National Park Service (NPS), and his goal was to make the parks accessible to the people. However, the NPS also was adamant that those roads must "lie lightly on the land." In other words, road design must consider aesthetics and maintain harmony with the surroundings. There was no sense getting people into the parks if the scenery they came for was desecrated.

From 1924 to 1933, the NPS and the Bureau of Public Roads (BPR) undertook the building of such a motor road—and not on easy terrain but over the mountains of Glacier National Park in Montana. Today the spectacular 52-mile Going-to-the-Sun Road hugs mountainsides, passes glacial lakes and arctic tundra, and crosses the Continental

135

Going-to-the-Sun Road in Glacier National Park
is a tribute to civil engineering and to aesthetics.

Divide at 6646 feet. Paved viewpoints allow the traveler to stop and enjoy the breathtaking views along the way.

It was impossible to travel across the park by auto before Going-to-the-Sun was built. Visitors took a train to what is now West Glacier, a stagecoach ride to Lake McDonald, and then an eight-mile boat trip to the Glacier Park Inn. Once there a visitor's only way to reach the surrounding one million acres of natural beauty was via horseback.

GOING TO THE SUN

The Going-to-the-Sun Road project was just the beginning of the "Golden Age" of park road construction, which began in the mid-1920s and continued until World War II brought it to a halt in the '40s. Yosemite, Mount Rainier, Grand Canyon, Rocky Mountain, Sequoia, Yellowstone, and Zion are some of the national parks that acquired paved roads during this era. In some cases, trails have been asphalted too, so hikers will keep to these prepared routes and avoid adverse impacts on the ecosystem throughout the parks.

BPR highway engineer Frank Kittredge remembered surveying the route over the Continental Divide for Going-to-the-Sun: "At the beginning of the survey every man not only had to walk several miles to work but also had to climb about 2700 feet vertically every morning before starting the actual work. This would be equivalent to climbing the Washington Monument five times before getting down to the job at hand every morning. . . . [It] was an undertaking to put even the mountain sheep to shame."

One observer has described how the $2.5-million road meets the NPS guidelines: "With masonry guardrails constructed from the very same rock excavated

In 1916, the hotel at Glacier National Park was only accessible by train, stagecoach, and boat.

for the roadbed, the Sun Road stands as a tribute not only to civil engineering, but to aesthetic regards as well. Retaining walls blend into the hills and its contours snake in and out of view, making Going-to-the-Sun one of the least obtrusive paved roads around."

This early cooperative effort of the NPS and BPR was so successful in allow-

A view looking down on Going-to-the-Sun Road from Logan Pass, the highest point of the road.

ing access while "lying lightly on the land" that it became the model for road construction in all the national parks.

In January 2004, Montana Congressman Denny Rehberg called the Going-to-the-Sun Road "one of the crown jewels of any highway within our park system." Since 1983, the road has been listed on the National Register of Historic Places, in 1985 it was named a National Historic Civil Engineering Landmark, and in 1997 it was designated a National Historic Landmark.

Teddy Roosevelt with John Muir at Yosemite. The view was much the same as the circa 1899 photograph below.

AUTOS COME TO YOSEMITE

On Saturday, June 23, 1900, Oliver Lippincott drove the first automobile into Yosemite Valley. It was a Locomobile, from the company owned by Asphalt Tycoon Amzi Barber. Lippincott had successfully reduced a stagecoach trip by half. Three days later, the second car arrived. Soon hundreds of thousands of automobiles would be entering Yosemite Valley each year.

In May of 1903, President Teddy Roosevelt visited the park. His host was naturalist John Muir, who later founded the Sierra Club. Muir used the view from Glacier Point to convince the President that Yosemite should be put under the protection of

Yosemite as it appeared in 1888. Until 1907 the only access was on horseback or by horse and wagon over rudimentary dirt roads.

the federal government. Four years later, the official in charge of Yosemite Valley banned automobiles, creating a public uproar, but the ban was not lifted for six years. On July 31, 1926, an "all-year highway" opened, helping to raise the total number of automobiles traveling into the park to 137,296 in 1927 alone. Nonetheless, that all-year highway was an unpaved road.

By the 1930s, Yosemite Park's roadways were suffering. Erosion and dust were killing the vegetation along the popular routes. Lives were at risk from the poorly graded and dangerously undermaintained mountain roads. In paving the roads, the federal government saw a win-win situation. One of the ways it found to put Americans back to work during the country's horrific economic depression and to preserve a national treasure at the same time was to pave some of Yosemite's roads.

ROCK SOLID

The Going-to-the-Sun Road in Glacier National Park was nearly ready for dedication when paving began in Yosemite. The challenge might not have been as extreme, but the same rules of lying lightly on the land would apply.

Starting his road-building career in 1929, the year of the stock market crash, was a young man in his early twenties named George Wagner. Wagner set out to make road-building history with his employer, Granite Construction Company, by building and paving the roads through Yosemite National Park.

In 1932, hearing that the company had won the contract in Yosemite, Wagner recalled, "Everyone was elated. Most of us were fishermen and hunters. We couldn't wait to get there. As soon as the snow melted and the roads opened up, Granite began moving in their equipment."

BUILDING A MONUMENT

The job was performed under contract to the National Recovery Administration (NRA), part of the New Deal alphabet soup. Under the NRA, the federal government funded public works projects to "make work." The practice was also called "priming the pump"—feeding money and jobs into the ailing economy in order to encourage growth. The employees' work hours were restricted in order to spread the work to more people.

There were two six-hour shifts, six days a week—from 6 a.m. to 12 noon and from noon to 6 p.m. "As I always worked the first shift, every afternoon was free. These afternoons were spent hiking, fishing, hunting when in season, and swimming. Trails were not always marked and were hard to find. We never got lost and almost never saw any other people on the trails. We hunted mountain quail and deer. We went to the dances on Saturday nights at Camp Curry. We always found something to do for entertainment," Wagner recalled.

Finding laborers during the Depression was easy, but finding skilled heavy-equipment operators was difficult. As bulldozers were a new type of equipment, operators were few and far between. Wagner and another man were the only skilled bulldozer operators on the job at first, and "We both spent a lot of time training operators."

"Priming the pump" was the federal government's strategy for feeding money and jobs into the ailing economy via vast public works projects during the Great Depression.

MAKING CAMP

Wagner was part of the crew sent to Yosemite early to make camp for nearly 100 men who arrived a week later. Downriver from the quarry and rock-crushing plant and a few miles from the popular Wawona Hotel, the campsites were furnished with a mess hall, offices, kitchens, maintenance shops, outhouses, and a trash-disposal site. "The garbage pit," George remembers, "was a haven for bears."

Best of all, "About 200 yards further downstream from the last tent in the bachelors' camp was the most beautiful swimming hole. That was where several of us took our daily baths—skinny-dipping."

ROAD PIONEERING

"We pioneered a new road up to Glacier Point in Yosemite. The old road was too rugged and too dangerous for automobiles. It was more of a trail than a road. There wasn't any pavement and the gradients were severe. The National Park people said it had to go. We built a new road and paved it. Pioneering the road was hazardous work. When you knocked a rock loose, it would roll 3000 feet down the side of the mountain," Wagner explained.

The first order of work on the Glacier Point Road job was clearing and removing the sugar pine, yellow pine, and ponderosa pine, which was highly marketable timber. Quarrying the stone was also part of the contract.

"We dug the roadbed down a foot and a half deep and about 20 feet wide into the face of the granite mountain. We were operating some of the first bulldozers ever used. The dozers were mounted on tractors. They were the biggest tractors built in those days. After the drilling and blasting, the rock would be pushed downhill into the crusher," Wagner reminisced.

George Wagner at the controls of a 60 horsepower "Cat" on the Glacier Point job at Yosemite.

The stone was crushed and screened, then trucked to the roadbed where crews spread it by hand. A water truck that pulled water from nearby creeks followed alongside the grader to reduce the dust and help compact the roadbed for rolling. An iron three-wheeled roller was used for compacting.

Finally, after having cleared the trees, blasted the right of way, quarried the rock, graded the roads, and placed the base, the crews would put down the asphalt. The thickness depended on how much traffic the government projected would be traveling that particular route.

The crew would place six inches of rock, then apply a penetration coat of hot asphalt cement. The first layer was made up of stones as large as two and a half inches in diameter. Next, three or four layers of rock, each about four inches thick, were placed. Each layer contained a progressively smaller stone—first one and a quarter inches, then three quarters of an inch, then one eighth of an inch. "Between each layer an application of hot asphalt was placed down and covered

with a chip seal, a three-eighths-inch broken stone, and then rolled," recalled George. "We called the pavements 'built-up macadam.' For this type of construction, Granite Construction was the expert.

"The hot asphalt provided a waterproof coating and acted as a binder, holding in the chip seal. Broken stones, called screenings, assured interlocking action. The screenings were smoothed with a six-foot-wide drag broom followed by hand brooms and rolled with a tandem roller. The roller weighed 14 tons. The first roll would be made at a 45-degree angle to the road's shoulder, and the second pass at 90 degrees to the shoulder," remembered George.

Today, thanks to the road Wagner helped to pioneer and construct, visitors to Yosemite can drive to the Glacier Point overlook. There, at an elevation of 7274 feet, they enjoy a magnificent view of Yosemite Valley 4000 feet below and of the renowned rock formation known as Half Dome.

IN HARMONY WITH NATURE

Many roads in the National Parks started out as trails, then were improved with penetration macadam or chip seals, and eventually were paved with hot-mix asphalt.

Like the road to Glacier Point, many roads in the National Parks started out as trails, then were improved with penetration macadam or chip seals, and eventually were paved with hot-mix asphalt. Sometimes the pace of road improvements in the Parks is painfully slow, mainly due to funding. The paving of Going-to-the-Sun Road, for example, started in 1938, but was interrupted by World War II and did not resume until the 1950s. Development of the 450-mile Natchez Trace in Tennessee, Alabama, and Mississippi took from 1937 until 2005.

Some roads in the National Parks are treated as cultural resources because of their age and historical significance. On these roads, maintenance is not just a matter of keeping the pavement smooth and safe; maintaining the original look of the road is also important. In the early days of paving in the parks of the desert Southwest and California, the rock for building the road was actually quarried in the park so that that the color of the road would blend with the landscape.

An asphalt road in the Grand Teton National Park, Wyoming, in the 1930s.

Quarrying is no longer allowed in most parks, making it necessary to locate similarly colored aggregate when these roads are rehabilitated, so that the asphalt continues to harmonize with the natural terrain.

Today the National Park System has approximately 8300 miles of roads, of which 5300 miles are paved, and more than 5000 of those miles are paved with hot-mix asphalt. Going-to-the-Sun Road, for example, is now all hot-mix asphalt, with the exception of the bridges and cantilever walls. The Blue Ridge Parkway, the Natchez Trace, Glacier Point Road—all are hot mix.

CHAPTER TEN

Asphalting the Mother Road: Route 66

The highway to San Juan de la Cruz was a blacktop road. In the twenties hundreds of miles of concrete highway had been laid down in California, and people had sat back and said, "There, that's permanent. That will last as long as the Roman roads and longer, because no grass can grow up through the concrete to break it." But it wasn't so. The rubber-shod trucks, the pounding automobiles, beat the concrete, and after a while the life went out of it and it began to crumble. Then a side broke off and a hole crushed through and a crack developed and a little ice in the winter spread the crack, so the resisting concrete could not stand the beating of rubber and broke down. Then the county maintenance crews poured tar into the cracks to keep the water out, and that didn't work, and finally they capped the roads with an asphalt and gravel mixture. That did survive, because it offered no stern face to the pounding tires. It gave a little and came back a little. It softened in the summer and hardened in the winter. And gradually all the roads were capped with shining black that looked silver in the distance.

—John Steinbeck, *The Wayward Bus*

Route 66 was one of the first highways where Americans could drive for miles on asphalt.

WHEN ROUTE 66 was created in 1926, the "highway" traveled 2448 miles through eight states, including the three youngest, Arizona, New Mexico, and Oklahoma. Running from Chicago to Los Angeles, the road was one of the first major interstate routes to be paved along its entire length when its last patch of road was asphalted in 1938. What started out as a dirt road through Missouri, Kansas, Oklahoma, Texas, New Mexico, and Arizona was book-ended by preexisting concrete sections in California and Illinois. Over the years, it became a mostly asphalt highway along its western half. After World War II, many miles of its concrete sections would be given an asphalt overlay or even ripped up and replaced entirely with asphalt. The famous road always had a rich, strong, and ever-deepening tie to asphalt.

California highway officials began building many of their state's highways with concrete, hoping they would last for decades. Unfortunately, they did not. So while the other Route 66 states began paving their first sections of the interstate highway, California, in the words of John Steinbeck, "capped the roads with an asphalt and gravel mixture."

NAMES TO NUMBERS

Route 66 was created from a random collection of dirt trails and roads that were strung together by five men acting on behalf of the 48 states and the federal government. As the first significant federally funded, nationwide road-building effort was in full swing, the United States was about to experience a radical shift along its highways. The chief of the Bureau of Public Roads, Thomas H. MacDonald, announced that the approximately 450 named trails and their related associations were to be no more. Which meant that the Lincoln Highway, the Jefferson Highway, and the fabled National Old Trails Road would carry numbers, not names.

Part of the National Old Trails Road was to become Route 66, which would earn the nickname "the Mother Road." Author John Steinbeck coined the term in his legendary novel *The Grapes of Wrath*. The book enlightened the public on the plight and the path of hundreds of thousands who fled starvation and homelessness. During the Great Depression, the Mother Road provided for the needy, nourished the starving, and carried the nation's poorest through the darkest of times.

Steinbeck wrote, "Highway 66 is the main migrant road . . . 66 is the path of a people in flight, refugees from dust and shrinking land, from the thunder of trac-

A section of the 1912 coast-to-coast National Old Trails Road would become part of the new Route 66 when numbers were substituted for names.

Left:
Families by the thousands loaded their possessions on their cars and fled the drought-stricken Dust Bowl via Route 66 in the 1930s. Below: Steinbeck's *The Wayward Bus,* as quoted in this chapter's opener, memorably describes asphalt versus concrete paving.

Left:
Families by the thousands loaded their possessions on their cars and fled the drought-stricken Dust Bowl via Route 66 in the 1930s. Below: Steinbeck's *The Wayward Bus,* as quoted in this chapter's opener, memorably describes asphalt versus concrete paving.

tors and shrinking ownership, from the desert's slow northward invasion, from the twisting winds that howl up out of Texas, from the floods that bring no richness to the land and steal what little richness is there. From all of these the people are in flight, and they come into 66 from the tributary side roads, from the wagon tracks and the rutted country roads. 66 is the mother road, the road of flight."

Physically and symbolically, Route 66 mirrored the spirit of the nation. When economic recovery came, it was coupled with a massive mobilization for World War II. Before heading to Europe, General George S. Patton conducted military exercises with live ammunition along the highway in California. Airstrips, many of which were paved with asphalt, were built along Route 66 for training fighter pilots. Throughout the war, the newly paved road carried its domestic armies of civilians to well-paying military manufacturing jobs.

TAKING ASPHALT TO THE HINTERLANDS

Romans built the first legendary network of roads. Almost 2000 years later, the British revived the art and science of road building with their macadam surfaces. The Parisians placed the first bituminous surface on a major city's grand avenue.

The United States, however, was the first country to aggressively pave its rural highways with asphalt. During the 1920s and 1930s, "The miracle was not the automobile. The miracle of the early twentieth century was the construction of a vast network of highways that gave the automobile someplace to go," explained Route 66 historian Susan Kelly.

"Not many people traveled in the early twenties. The roads were no good, tires were no good, cars would get hot, and the highways were not marked. People would stop here and ask, and they wouldn't know if a town were in Kansas or New York." Frustrations were numerous among the unmarked and unpaved roads, recalled Michael Burns, who as a young man witnessed the confusion of motorists who stopped at his father's filling station on Route 66 in Springfield, Illinois. "One day there were two fellows who came in here and asked if this was Springfield. I said. 'Yes,' and they told me to go ahead and service their car. When I told them it came to fourteen gallons of gas and two gallons of heavy oil, they were stunned. 'My golly!' They said, 'We left home this morning and we was only forty or fifty miles from Springfield.' They had turned wrong and had come to Springfield, Illinois, instead of Springfield, Missouri. That happened a lot in those days. A lot of people turned wrong."

As the 20th century dawned in the United States, there was no such thing as a highway that was numbered from coast to coast. There was not even a paved highway that ran between two major cities. Consistently signed and well-paved highways simply did not exist—not one. Some roads were gravel, others were compacted clay or even crushed seashells, but none of them was asphalted along its entire course. The nation was lusting for dustless, mudless roads, the kind that could be driven on no matter how much rain or snow fell—true all-weather roads.

"It was not Highway 66 at the time. It was not even a road," said Kelly. Storming out of Los Angeles on a motorcycle in the early 1920s, with his bride clinging to his back, George Greider set out for the old Cajon Pass, hoping to find a 19th-century military wagon road called the Old Beale Wagon Route. The pavement out of Los Angeles was only wide enough for a single vehicle and, as Greider explained, "We really moved on that slab of paving." But all good things must come to an end. "There were no signs—nothing—and we just drove off into the sand. The sand was real deep, and the motorcycle came to a complete stop. . . . We stood there a while. There were two ruts in the sand from there on . . . for miles and miles, hundreds of miles. There was no sign of civilization at all: no road work, no fences, no grading, no highway."

A **1920s** wooden-plank road in the California desert shows one effort at paving difficult areas.

Before it was asphalted, a small portion of the road that became Highway 66 was "improved" with wooden planks. Richard Warren, a road warrior who drove on the plank road as a child, explained the rules of engagement when bouncing along the lumber highway: "If two automobiles met, each kept one set of wheels on the planks so as not to get all four of either vehicle mired in the sand."

The monumental effort of linking the nation's cities together by paving its rural highways began after World War I. With MacDonald, the "Chief," at the helm of the Bureau of Public Roads, a diverse group of associations, manufacturers, road builders, and most important, the heads of each state's highway departments, began working toward the previously unthinkable: a system of interstate highways. In 1924, the Chief's goal at first was modest: "My aim is this: We will be able to drive out of any county seat in the United States at thirty-five miles an hour and drive into any other county seat—and never crack a spring."

By this time, the notion of paving a city street with asphalt was no longer a radical one. But many thought that covering thousands of miles of dirt roads in the hinterlands and beyond was overly ambitious. Even the believers weren't too sure how it was going to be done. One thing everyone was sure of: Life in the United States got better with each mile of highway that was paved. Americans were starting to move about the country at will, not just when the weather or the train schedule permitted. Pavement was liberating Americans and improving their lives forever.

Thomas H. MacDonald, the chief of the Bureau of Public Roads

Ads for the Barber company extolled lake asphalt, but asphalt refined from crude oil prevailed by the 1920s.

ASPHALT, THE PEOPLE'S CHOICE

In the 1920s, the Barber Asphalt Paving Company and the Warren Brothers were still giants in the asphalt paving business, but much was changing. Lake asphalt no longer dominated the industry, as asphalt refined from oil had become the standard. The patents on plants and pavement mixtures began to expire, allowing small contractors and their customers to create and use a hot mix with any size aggregate they wished without fear of being sued for patent infringement. Federal and state engineers were making public the findings of their own research and development. Road-building equipment such as portable plants, rollers, and even trucks—previously out of reach for small, independent contractors—became affordable.

Entering the contracting business was an attractive proposition for the ambitious in the 1920s. Construction, of both roads and structures, was everywhere, and the capital required for starting a contracting business was relatively low. Many energetic young men—especially immigrants and members of minority groups, including Italians, Irish, and Eastern Europeans—were drawn to contracting.

James Francis Gallagher was one of these young men. Jim had been an office boy, running messages for the Barber Asphalt Paving Company in Chicago. Answering the call to serve his country during World War I, he left his office job and was accepted into the United States Navy. As a midshipman on an ore boat plying the Great Lakes, he and his crew ferried iron ore from Minnesota's Masabe Range to factories in Gary, Indiana, and the South Side of Chicago. After peace was declared, the young seaman landed a job selling asphalt for Texaco.

Texaco produced an asphalt crack filler that was used on Illinois's fairly new but badly beaten-up concrete highways, as well as the millions of miles of cracks between the millions of bricks in the streets of Chicago and its suburbs. Texaco kept Jim busy until an offer with more promise came in the form of another job, this one with the Rock Road Construction Company. The owner of the firm, Sam Nannini, an Italian immigrant who didn't speak English well enough to communicate effectively with municipalities regarding asphalt roads, hired the persuasive Jim Gallagher to do so on his behalf.

Much was at stake. If a town's government could be swayed to vote in favor of improving its streets with asphalt pavements, that town's administrators might contract with Rock Road Construction for the work. With his mission clear, Jim set out in his company car. Talking asphalt roads with the locals was a natural task for the gifted salesman.

Road-building equipment such as portable plants, rollers, and even trucks—previously out of reach for small, independent contractors—became affordable.

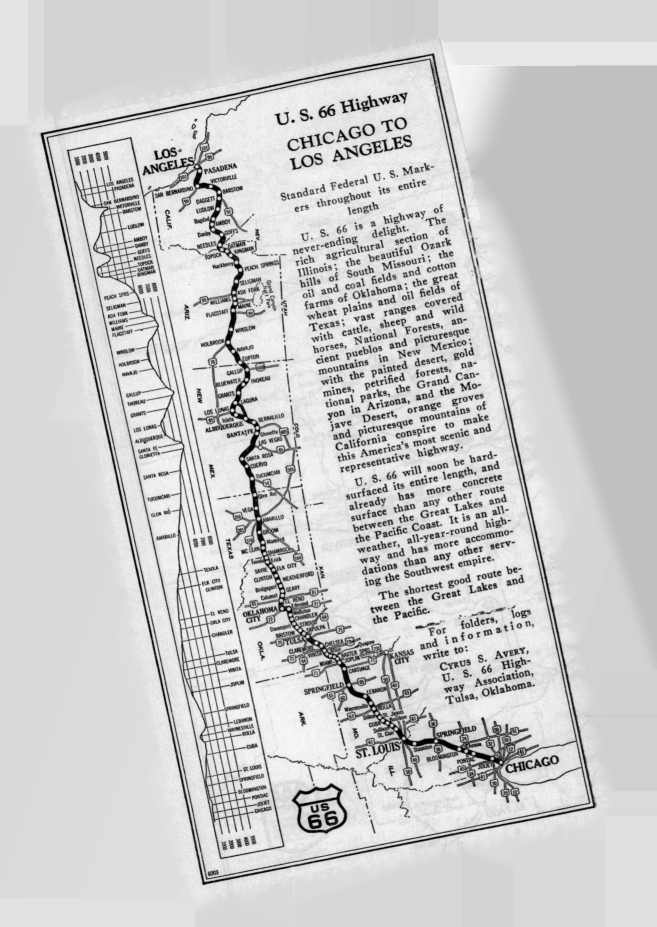

U. S. 66 Highway

CHICAGO TO LOS ANGELES

Standard Federal U. S. Markers throughout its entire length

U. S. 66 is a highway of never-ending delight. The rich agricultural section of Illinois; the beautiful Ozark hills of South Missouri; the oil and coal fields and cotton farms of Oklahoma; the great wheat plains and oil fields of Texas; vast ranges covered with cattle, sheep and wild horses, National Forests, ancient pueblos and picturesque mountains in New Mexico; with the painted desert, gold mines, petrified forests, national parks, the Grand Canyon in Arizona, and the Mojave Desert, orange groves and picturesque mountains of California conspire to make this America's most scenic and representative highway.

U. S. 66 will soon be hard-surfaced its entire length, and already has more concrete surface than any other route between the Great Lakes and the Pacific Coast. It is an all-weather, all-year-round highway and has more accommodations than any other serving the Southwest empire.

The shortest good route between the Great Lakes and the Pacific.

For folders, logs and information, write to:

CYRUS S. AVERY, U. S. 66 Highway Association, Tulsa, Oklahoma.

"Sam Nannini wanted him to drive in style and insisted that his company car be a Packard," explains Jim's grandson, Dan Gallagher. "My grandfather wasn't comfortable with such a high-profile car when he was talking with people of fairly modest means, but Sam insisted. So he would head out to a suburb, set up a keg of beer on a street corner, tap the barrel, and start pouring beers. As people came to have a drink, my grandfather talked to them about the cost and benefits of asphalting their roads."

As much as the men would like to have believed it was their decision alone on whether to pave, another impetus for road building often came from the women of the house. "More often than not, my grandfather said, it was the housewives who demanded that the roads be paved. They were sick and tired of cleaning up after the mud and dust that came in the door every day. The women were insisting on paving the roads!" The public demanded new roads, and town governments were persuaded to build them.

Many miles of local roads in Chicago were soon paved with asphalt, the material of choice for most city pavements in the 1920s. In the 12 largest cities in the United States, asphalt streets made up more than half of all paved surfaces—more than all the other forms of paved roads combined. Concrete streets made up less than three percent of paved surfaces.

The asphalt industry was maturing. Jim Gallagher went out on his own and formed Gallagher Asphalt Corporation. Becoming the executive secretary of the Association of Municipal Contractors of Chicago, Jim branched out, making more contacts and becoming a major contractor in the southern suburbs of Chicago. Today Jim's grandchildren including Dan still carry on the family business, and it is still helping to build Chicago.

AMERICA'S FIRST INTERSTATE HIGHWAYS

"Route 66 is a giant chute down which everything loose in this county is sliding into southern California," remarked Frank Lloyd Wright, the famous architect whose work graced Chicago and the Midwest.

Not far from where Gallagher Asphalt eventually set up offices was the zero mile of a new highway that helped bring about a new era in America. U.S. Route 66 began near the edge of Lake Michigan in Chicago's Grant Park. Driving from east to west, the intrepid travelers who dared take the highway to Los Angeles made their way out of the hustle and bustle down Jackson Boulevard, passing through Al Capone's turf in Cicero, past the farm fields of Illinois, and eventual-

Opposite:
Route 66 was "a highway of never-ending delight" and "the shortest good route between the Great Lakes and the Pacific," said the American Automobile Association.

Cyrus Stevens Avery tirelessly promoted the interstate highway, earning him the title of Father of Route 66.

ly to St. Louis, where the excitement of the open road grew after turning westward. From the Ozarks on, a driver was enchanted by dairy farms, herds of roaming cattle, desolate Indian lands, the Black Mountains, the Mojave Desert, and finally the land of milk and honey nestled next to the Pacific Ocean.

Every aspect of the highway was unusual. It ran not east and west or north and south but at a diagonal. It was not the path of one established trade route or named trail, but rather a new road following the routes of several. The architect of its course was another highway entrepreneur and visionary, Cyrus Stevens Avery.

As early as 1913, Avery had been inspiring Oklahomans to take charge of their roads. A prominent citizen of Tulsa and head of the region's Ozark Trail organization, he gathered thousands at rallies, convincing them to improve the dirt trails through Oklahoma. This was no small feat in a rugged pioneering land that was not even a state until 1907. The roads were not marked or maintained, and as a result they were hazardous to travelers. Making matters more difficult for road supporters, trains were safer and easier to use.

Long before others understood the potential commercial might of modern highways, Avery was promoting their construction. He became the father of Route 66 by imagining a highway through the Southwest that linked Chicago and the West Coast. As a real estate speculator and an entrepreneur with both a car dealership and one of the first tourist courts in the state, Avery let his business acumen and instincts guide his drive to see the road built. And as a member of numerous road associations, he used his influence.

Avery secured state funding to have split-log drags, a primitive but effective tool to smooth the dirt roads, delivered to the farmers living along major travel routes. He was then able to offer a dollar a mile to the farmers dragging the split logs down the byways close to their farms. The ideal time for this work was after mild rains. His plan became the first significant road-improvement program in Oklahoma. As a result of Avery's efforts, in 1917 over 150 miles of roads were being dragged and made passable within 24 hours of a rainstorm.

Appointed Oklahoma's state highway commissioner in 1923, the forward-thinking Avery proposed a state-imposed gas tax as a way to raise money for road improvements. But it was his actions over the next few years that led to the creation of a U.S. numbered highway, Route 66, right through his home state of

Oklahoma and the heart of the Southwestern United States.

In 1924, as the state's head of highways, Avery became part of an elite group of his peers, the American Association of State Highway Officials (AASHO). The 48 heads of the state's highway departments and their immediate staff made up its membership. In cooperation with the Chief, MacDonald, AASHO set out to identify a series of numbered highways whose numbers would replace the names of the trails. The federal-and-state joint venture created the nation's first numbering system for interstate highways, a grid crisscrossing the United States and known as the U.S. Routes. Americans, for the first time in the nation's history, could now follow an organized system of numbers to just about anywhere in the country. The even-numbered routes ran east–west, with the lowest number ending in 0 in the north and the highest, U.S. 90, in the south. Odd-numbered routes ran north–south, with the lowest, U.S. 1, on the East Coast and the highest, U.S. 101, in the West.

Before asphalt paved rural highways, split-log drags smoothed the dirt roads. Here, a car pulls one in the state of Washington, circa 1915, as a man adds his weight to it.

PAVING "THE MAIN STREET OF AMERICA"

Avery suggested to the newly formed U.S. Highway 66 Association that the highway be referred to as The Main Street of America. Using the name in paid ads and promotional materials, the booster group was prompting residents of the other 40 states to take a trip through the eight U.S. Route 66 states. The Main Street of America label created enthusiasm in the citizens of the towns the highway passed through. It was equally effective at directing the travelers and tourists driving it.

In at least one town, the locals were inspired to help in paving the road. Kent Ruth, a local historian from Geary, Oklahoma, said, "The amount of volunteer work that went into building up 66 in those days was amazing. Everybody had to work on the road. If a businessman couldn't work, he would hire a kid. That gave the high school boys a chance to get out of school for a day or so."

People living along the highway monitored the progress of the paving with excitement. The major dailies as well as the small-town newspapers provided detailed accounts of the work progressing along the highway. Crowds gathered for milestone events, drawn in by the promise of what the paved highway was going to deliver to them personally.

In 1927, John Woodruff, president of the U.S. Highway 66 Association, inspired a crowd of 500 gathered in Amarillo, Texas. Cheering on the paving in the Texas Panhandle and the rest of the highway, he proclaimed, "[I]f the development of the highway continues, I am sure it will be entirely paved by the end of 1928." Excitement was clouding Woodruff's judgment—he was 10 years off in his prediction. Nonetheless, U.S. Route 66 would be one of the first major U.S. routes to be completely paved.

The lack of pavement on some stretches of the highway created a hardship to travel that some were not willing to endure. Russell Byrd recalls his first trip as a bus driver along the early sections of the highway through the Mojave Desert. "The first time I went into Needles off

The AAA has been producing maps for motorists since 1905. This one of Route 66 shows the segment from Needles to Los Angeles, California.

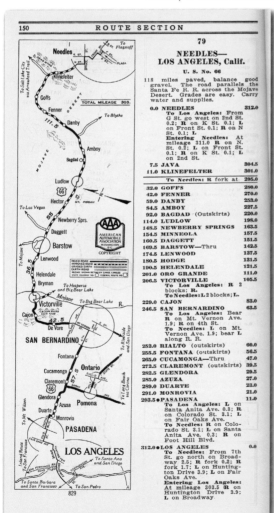

the desert, I looked back in the bus and couldn't recognize the passengers. They were all covered with dust. Later one of the drivers told me how to regulate the air by opening up the wing windows," but it was too late for those on his maiden voyage.

If dust wasn't plaguing passengers, mud was. Bob Lee grew up working for his father's bus company, collecting dimes from passengers renting pillows on the family's 11-passenger buses during the long trips up and down Route 66. "In those days Highway 66, as it is now known, was a very difficult road to travel, particularly when it rained. It was just mud and clay mud," recalled Lee. Speaking about a stretch of the highway between Geary and Clinton, Oklahoma, Lee remembered the days before pavement: "There was a real mud hill at Geary. If it was raining, the buses wouldn't leave Clinton [about 40 miles west] because they couldn't get up the hill at Geary. A lot of buses got stuck in the mud there, and they had to put the people up in hotels until the buses could be pulled out."

On July 16, 1932, the first national ad of the U.S. Highway 66 Association ran in the *Saturday Evening Post*. Costing $2000, it was a major expenditure with a provocative suggestion: Drive yourself to the Los Angeles Olympics along the Great Diagonal Highway. The point being made was, "You stay on pavement longest going west. You get on the pavement first going east." Over 700 people sent in the attached coupon requesting more information.

Heading west to start a new life created a migration along Route 66 and a population boom in the West. Between 1920 and 1930, the population of Los Angeles doubled. Many of the city's new residents arrived by way of the Mother Road.

Phillips 66 gasoline's name played on public enthusiasm for the route. A special executive committee had been appointed to determine a trademark. The company's first refinery was near Route 66, and when a company car testing the new Phillips gasoline zoomed 66 miles an hour down highway 66, it all came together and Phillips 66 was approved unanimously.

MAKING DO WITH CONCRETE EQUIPMENT

The total miles of asphalt roads far exceeded the total miles of concrete highways. Regardless, the early concrete pavement contractors had equipment designed and built exclusively for their needs. Well into the 1920s and '30s, however, asphalt contracting was a job most often performed by hand. If an asphalt contractor wanted to use equipment to lay down a hot mix, he had to retrofit concrete pavers geared for the cold mixing of concrete.

Portland cement concrete was laid by automated pavers moving slowly over rails and tracks made of wood, called sideforms. The paver left a continuous slab of roadway in its path, but the sideforms had to be removed and replaced as the paver advanced.

An asphalt drum mixer reveals its origins in the concrete industry, source of much early asphalt equipment.

In 1927, the Griffith Paving Company may have performed the first hot-mix asphalt paving jobs ever to use an automated paver, while working on a 16-mile section of highway in Orange County, California. Taking a concrete paver and using it on an asphalt job was a gamble that paid off. County engineer Nat Neff, who was overseeing Griffith's work, found that the asphalt pavement had greater consistency, could be rolled faster, and was smoother and more uniformly dense. The entire job, which had been estimated to take five months, was completed in just three.

About the same time, in Oklahoma, H. L. Cannady Co. was having a hard time hiring skilled and reliable asphalt rakers. C. L. Donovan of the Bureau of Public Roads suggested that the Cannady Co. use a concrete finisher that was still on the job site from having laid the concrete base of the road. G. H. James of the Oklahoma State Highway Department agreed to the impromptu field trial. The results were impressive—the surface had excellent consistency in density and smoothness. The only adjustment to the equipment was that it rode along concrete headers instead of timber or steel, and the screeds that leveled the asphalt were heated with torches before operation.

With automated pavers, contractors achieved a 50 percent increase in the speed of laydown. The paver could pave faster than the batch plant could make hot mix. The raking problem had been resolved, and perhaps more significantly, highly skilled shovelers were no longer important. Up until this point, shovelers had to keep their eyes on the grade that the design required; now they just needed to feed the paver with a steady pile of hot mix. Automated asphalt pavers, even those that had originally been designed for concrete, eliminated the need for crews to double back on their own work to address and improve rough spots.

Similar discoveries were also made in Tennessee and in both North and South Carolina, but it was in California that the results were so evident that California's Division of Highways became the first state highway department to require mechanical pavers on all state jobs.

Automation in laydown was changing everything. *Engineering News Record* commented in 1927, "Spreading and raking of asphaltic concrete in road work has almost always been a hand job." Using two concrete spreaders, an asphalt contractor was able to break new ground by laying down 43,556 tons of hot mix between August 18 and October 8, averaging 1050 tons per eight-hour day. "With virtually only two years of considerable use, machine finishing has established itself not only as practicable but as a money- and time-saving method of performing mechanically about the only remaining process in asphalt paving," reemphasized the magazine just a year later. There was no going back. The day of automation had arrived for the world of the asphalt pavers.

In 1928, on a job in California, 820 tons of hot mix was put down in one day—an amount thought to be impossible by hand. It was clear: Automation meant the pavement was smoother, more uniform, and less expensive to build. There were fewer skilled shovelers and rakers required, meaning there was less expense to the contractor and the contracting agency. Automated jobs required less inspection. They also consumed far less fuel, partly because backpatching—the heretofore routine process by which a crew doubled back to patch their work—was not necessary. Contractors even found that they could deliver asphalt to the job site at a cooler temperature because instead of being left standing before being placed on the road surface, it was laid down right away.

Automated asphalt pavers eliminated the need for crews to double back on their own work to address and improve rough spots.

AN ASPHALT REVOLUTION ON ROUTE 66

"If it weren't for the floating screed we'd all be pouring concrete. It was that important an invention. Without it the asphalt contractors would've been dead," explained David Emerson, a contemporary expert on asphalt equipment. The floating screed changed the asphalt contractor's world forever by mechanizing the laydown. The invention has been called the most significant and important development in the history of the industry.

The 1930 Road Show, an annual event showcasing the road-building industry's newest road-building gear, was held in Atlantic City, New Jersey. The show was notable for what it *wasn't* showing as much as for what it *was* featuring. It was the first year steam shovels would not appear, as they were replaced by diesel-powered shovels. It was also the first year the Barber-Greene Company demonstrated its floating screed.

Harry Barber and William Greene were pioneers in asphalt equipment who introduced "the first functional bituminous concrete paver." In 1916, seeing the federal and state governments' powerful commitment to funding and paving roads, the two men had founded the Barber-Greene Company. Barber was an inventive genius who was eventually awarded 70 patents. Greene was a master salesman. He provided the organizational power that was critical in getting road builders to accept the new devices that Barber developed, while structuring the firm to be fiscally sound and competitive. At first they made bucket loaders and belt conveyors, and eventually became one of the leading manufacturers of road-building equipment.

At the conclusion of the Road Show, the experimental floating screed was shipped directly to the job site of Sheldon G. Hayes, an asphalt contractor headquartered in the Route 66 town of Rolla, Missouri. It created a buzz as word spread quickly that Hayes was using Barber-Greene's experimental device designed just for hot mixes, making Rolla the first town ever to employ such a device. The contraption was met with skepticism as it was still in the research and development stages. Regardless, Hayes was eager to put it to use and see if it could successfully take a hot asphalt mix and apply it to the road.

The genius of the floating screed was that it did not need side forms. Barber-Greene eliminated them by using a sliding dolly that allowed the screed to "float." The floating screed, passing above the surface of the road, automatically placed more asphalt into low spots and less in the higher areas of the road base, further eliminating the need for shoveling more asphalt into low spots and scraping down high spots.

The first paving job ever done by a Barber-Greene asphalt machine was in Rolla, Missouri, in 1931. The 20-foot-wide finisher operated on rails and was towed and fed by the combination bucket loader and mixer. Inset: the original patent drawing of the bucket loader.

In 1932, gambling the company's future while the country was reeling in the depths of the Great Depression, Barber-Greene continued pouring resources into developing the experimental device. The goal was to get all the tasks previously performed by hand to be performed in a single pass of a single machine. In 1934, Barber-Greene called its floating screed the Model 79. Hot mix was collected in the Model 79's hopper from a dump truck moving in front of it. "Flight feeders" then fed it onto augers that churned it and force-fed it to the finisher, which was equipped with tamping screeds. These spread the mix into a compacted two- to three-inch-deep travel surface. Moving on tractor treads, the machine created a swath of finished pavement 10 feet wide.

The Model 79 had its limitations at first. Cold mixes and hot mixes with medium and heavy aggregates could be used successfully, but finer aggregates gave the machine trouble. One observer remarked that it laid sheet asphalt only after tearing up the existing surface.

By 1936, Barber-Greene could boast that it had a far better paver.

By 1936, Barber-Greene could boast that it had a far better paver. Touting it as the Model 879, the company put the machine into production, confident that it could handle both heavy and fine pavement mixes. The machine had improved leveling ability and a highly effective tamping screed that finished previously difficult-to-lay sheet asphalts into a smooth surface.

Not all the credit for the idea of the floating screed could go to Barber-Greene, as two other firms, Adnun and Jaeger, had developed their own machines that did not require side forms. But it was Barber-Greene that achieved many important firsts: the first transverse distributing augers, the first crawling tractor treads on a paver, the first tamping screed, and the first flight feeders. Barber-Greene enjoyed the greatest success of the industry's most critical step forward. And it all started with a test run in a Route 66 town.

ASPHALT ROADS BRING BETTER ECONOMICS

In the throes of the Great Depression, road building sometimes meant salvation. To desperate men seeking a job, paving a road could mean food and shelter for their families. To the rest of the nation, road building became a way to revive the economy by "priming the pump." Road programs became a hallmark of President Franklin Delano Roosevelt's Works Progress Administration (WPA) and a key tool for economic recovery.

On the Great Plains in the 1930s, the once rich topsoil began turning to dust. The dried soil whirling in the strong winds turned into ferocious dust storms. On May 10, 1934, a single dust storm stripped 300 million tons of topsoil away from Arkansas, Colorado, Kansas, Oklahoma, and Texas. Lester Dill, a pioneer in building roadside attractions along Route 66, said, "Those years of 1934 and 1936 were hot. The dust came from Oklahoma and Texas. It was all over the cars and came in the windows. You could smell it. Those were hot, dry years, and that's when people just packed up and began to leave."

The young highway became the Mother Road to farmers who lost everything to the Dust Bowl. It was as if the highway itself nurtured them and provided them with a path to a better life. If they were lucky, the displaced and homeless might even make a living paving it. "Dad, he built roads. He worked on 66 clear across the state of Oklahoma," reflected Grady Bell, whose father and family nearly "starved to death back during the Depression." Bell added, "We moved to town off the farm, and Dad worked for the WPA. He built roads. He hauled dirt with a team and dirt shovel, and he done that from about 1930 to 1938. That road was

DUST STORM APPROACHING SPEARMAN, TEXAS.
APRIL 14. 1935

built with teams and wagons."

"The Grapes of Wrath went through here in the thirties, all these Arkansawyers going to California to pick grapes. They had everything with them: pigs in a cart, chickens in a crate, and a dog on top," remembered Marita Bumpers, who, with her husband, ran a grocery store and gas pumps in the Texas Panhandle.

Unlike the railroads, whose patrons were captive to services at the stations where they embarked or disembarked, the highways generated revenues along the entire route and among the entire population. Wherever a car or truck could

The dust storm on Black Sunday in 1935 brought such darkness that people got lost in their own front yards or huddled in their homes fearing they'd smother.

stop, a buck might be earned. Serving drivers gave the road a needed boost.

The more fortunate scraped together a little money and opened mom-and-pop businesses along the newly paved highway. In 1937, a Route 66 construction crew took up residence in front of Homer Ehresman's restaurant on the Texas-New Mexico state line, providing some badly needed customers. "When they paved this road out here, we was feedin' three times a day, family style, all you could eat for a dollar a day. When they was building this highway they was doing it all by hand—they didn't have any of this modern equipment—and those people were getting thirty cents an hour and only worked forty hours a week. They slept out along the roads in tents. Seems to me that lasted all through the summer and into the fall that year—not too long," recalled Homer. Once the road was paved, the workers

"Okies," fleeing Oklahoma's Dust Bowl, took Route 66 to California. Farmers became migratory workers who followed crops as they were ready to harvest.

moved on and interstate truckers took up their seats in the new restaurant.

Glenn Johnson of Kingman, Arizona, recalled, "When the Okies got here, the merchants had an agreement. They would give 'em enough gas to get to California. What happened was that the merchants would fill the gas tanks to get the Okies to Needles, and then the station owners in Needles did the same thing—and gave them enough gas to get them to San Bernardino, then once they got there they would scatter. They were all good, hard-working people. Not any bums among them. A few stayed here and worked in the mines and became leading citizens."

On August 15–18, 1938, in Amarillo, Texas, a gala celebration was held for all those who loved Will Rogers and U.S. Highway 66. The highway's last stretch of dirt road had been paved a few months earlier with the completion of Federal Aid Project 643. The final paving job was small, an 18-mile section of asphalt on Route 66 between Adrian and Glenrio, Texas, but the significance was profound.

Americans could now travel one of its new interstate highways on pavement. The event memorialized Will Rogers, one of the nation's most beloved entertainers, who had died in a plane crash in Barrow, Alaska, three years previously. Each of the eight Route 66 states agreed to call their portions of the interstate route the Will Rogers Highway. The road had one number but several names—Main Street of America, Mother Road, and Will Rogers Highway.

MIXING, HAULING, AND LAYING DOWN

With hand tools and mules dominating the industry into the 1920s, and demands for paved roads going unmet, change to a mechanized and motorized asphalt-paving industry could not come soon enough. Paving jobs were getting bigger, requiring hauling equipment to jobs between cities, not in them. Now, with funding from the Bureau of Public Roads and the state highway departments, contracts were being let for projects to pave many miles of rural highways.

Will Rogers, popular humorist and entertainer, is memorialized in one of the nicknames for Route 66, the Will Rogers Highway.

Early road-building equipment was originally designed for farmers and concrete contractors, not asphalt contractors. Farm tractors were the first with continuous caterpillar treads, and giant paving trains and drum mixers were manufactured for the concrete industry. Not until Barber-Greene's floating screed in the 1930s was there an asphalt paver that completely automated the process of laying hot-mix asphalt.

Developments both minor and major altered the road-building and asphalt-paving industry. The U.S. Army, enamored with trucks, contracted with a company called Thew to produce the first truck crane in 1918. Five years later, road builders got a machine that offered the "strength of 12 men" when LaPlant-Choate started equipping tractors with steel shields—the first bulldozers. It was in 1932 that the now familiar highly visible yellow paint was first used to identify heavy equipment working on highways. Beginning a trend, Caterpillar Corporation abandoned their color scheme of gray with red trim and began painting their rigs "Hi-way Yellow."

Filling the gap between dirt roads and asphalt surfaces were roads with bituminous macadam and penetration macadam surfaces. In the rush to dustless roads, the only way most rural areas were going to get an asphalt coating on their highways was by a bituminous distributor truck. Large tanker trucks sprayed asphalt on old macadam roads. The liquid asphalt, it was hoped, would penetrate the macadam surface, coating the stones and locking in dust that would otherwise become airborne.

Trucks with 1000-gallon tanks were equipped with 20-foot-wide spray bars. In one pass, they could lay down about two and one half gallons of bitumen per square yard of roadway.

Trucks with 1000-gallon tanks were equipped with 20-foot-wide spray bars. In one pass, they could lay down about two and one half gallons of bitumen per square yard of roadway. Over time, the amount of asphalt sprayed increased and the accuracy in which it was sprayed improved, allowing crews to avoid leaving "fat and lean" spots. Increased spraying widths, faster trucks, and quicker turn-around times when refilling tankers meant that smaller contractors with just a few trucks could apply over a million gallons of liquid bituminous dust palliatives in a busy year.

"Tarvia, Better Roads at Lower Costs" became the slogan for the Barrett Company. Like the Warren Brothers Company, Barrett started out in the roofing business in the mid-1800s before breaking into roadwork about the time Pennsylvania Avenue was getting its first asphalt pavement. Skilled at working with both tars and asphalt, the company began marketing its own dust palliative in 1906, calling it Tarvia. Through a successful public-awareness campaign for Good Roads, the firm heralded, "Tarvia Preserves Roads, Prevents Dust," making the product a household name and landing it in Webster's Dictionary.

Laying dust palliatives over old macadam surfaces prolonged the life of the road and kept dust to a minimum, but heavy rains and traffic washed and wore the oils away. As a result, rehabilitating a road by ripping it up became a popular method of road improvement.

MECHANIZING ASPHALT

In 1926, the first mix-in-place section of road was laid down in California with much success. Instead of creating the mix in a plant, it was done right on the road. Road graders and harrows, pulled behind treaded tractors, churned thoroughly graded gravel with slow-curing liquid asphalt and blended it into the roadway. The churned mix of gravel, earth, and asphalt was left in place, and traffic from passing vehicles compacted it. Occasionally, a roller would be used to finish the job.

Later, mixed-in-place techniques were enhanced by tractor-drawn multiple-blade drags that were capable of churning the earth several times in one passing. Also, self-propelled motor graders came to the market by the 1930s. They were

A Barber-Greene plant creates hot mix to fulfill America's growing demand for asphalt roadways.

faster and could control the depth of the blades' penetration. The mixed-in-place method stabilized a road's bed, but it was no substitute for an asphalt surface.

In 1924, the Butler tailgate spreader quickly became a fixture on road-paving jobs, especially in the southwestern United States. Before this, tailgate rock spreaders attached to the back of horse-drawn wagons were used for dispensing asphalt over the roadbed. The Butler spreader attached to the back of a dump truck with two large chains. Adjusting the controls, an operator could vary the position of the adjustable "wings," altering the width of spread of hot mix on the roadbed. Two three-foot-wide rolling belts allowed the heavy metal spreader to trail behind the dump truck that was feeding it. It was a primitive start to mechanical spreading and nearly eliminated shoveling. Raking, however, was still required.

Between World War I and World War II, hot-mix asphalt plants did not change much. Six major firms were manufacturing plants—Warren Brothers Company, Iowa Manufacturing Company, Barber-Greene Company, Standard Steel Corporation, Simplicity, Hetherington & Berner, and Cummer. The Warren Brothers boasted the largest plant in the world, one they had assembled in Brooklyn, New York. The behemoth supposedly produced 10,000 square yards of asphalt per day. Competitors claimed that it was just four separate plants producing 2500 square yards per day, combined into one megaplant.

Getting the tons of stone needed to make hot mix to the cold feeders was always a challenge, if not a backbreaking task. A collection of wheelbarrows, horses, mules, shovels, and a lot of dust were on the scene at the cold feed. Even when a plant had the benefit of a large steam shovel to supply the cold-feed elevators with gravel, moving the crane could be as difficult as moving the entire plant and always required too much time, labor, and money.

Lighter steel cranes powered by gasoline and diesel engines that rode on air-filled rubber tires did more to change the plant than the invention of the gob box. Improvements in metallurgy made cranes lighter and easier to move. Diesel engines became smaller and more dependable. Portable cranes on pneumatic tires replaced steam shovels mounted on flatbed railroad cars or tank treads, making them easier to relocate. Also, crews were able to break a plant down and haul it over the highway in separate pieces. Trailers equipped with pneumatic tires were built to haul individual components of the hot-mix plant down the highway, making the asphalt plant easier to assemble and disassemble as well as transport.

By the 1930s, maintaining paved roads became a pressing issue. The public was demanding more paved roads with better surfaces so they could drive faster. Highway officials needed to figure out how to keep paving new roads while upgrading previously paved roads, without breaking their budgets. When making the repairs, they had to work quickly and inexpensively to avoid angering voters. The Depression years meant highway departments had to get creative with limited funding. They turned to flamethrowers.

Throwing flames and heat on the road, surface heaters helped highway crews and contractors with patching and repairing bumpy sections of the road. With surface heaters, old asphalt could be warmed to a depth of a few inches in a few minutes or less. Whether the heaters were held in their hands, mounted on their trucks, or affixed to trailers, road crews laid down their fire and then cut out the damaged road with rakes, hoes, and shovels, replacing broken-down areas with a smooth patch of asphalt.

MOTHER ROAD TO MILITARY ROAD

General George Patton, preparing for his invasion of North Africa in World War II, drilled his forces along Route 66 in the Mojave Desert. Glenn Johnson, who grew up nearby, recalled both the damage the general's tanks inflicted and the inconvenience caused by the military vehicles. "They just tore up 66 between Needles and Barstow. If you were driving across, you'd have to wait, then they'd

General George Patton's Desert Training Corps deployed along Route 66 in the desert in 1942.

escort you in a convoy, ten or twenty miles at a time. By the time you got to the next stop, that bunch ahead of you would be leaving. If a guy broke down, that was just too bad. They'd push the car over to the sides; he'd get in with somebody else and go on. We were in a war and unless you were a V.I.P. you did what they said. Anyway, they didn't want you out there with live ammunition. Patton fought the African War there first."

War changed everything along the highway. Road-improvement programs unrelated to the mobilization came to a grinding halt. The government rationed gas and tourism dried up. Migrant workers were no longer drifting, as well-paying manufacturing jobs were available for the taking. The young men in the farming towns went overseas to fight, and their parents went to California to work in military factories. Military convoys and trucks shipping everything from airplane wings and tanks moved up and down the highway. Passenger buses carrying troops and visiting families ferried between towns with military bases.

Although it overburdened the simple two-lane highway, the war wouldn't wait for improvements. Route 66 and the people living and working along it did their best to adjust as the highway became home to airstrips, army bases, training camps, prisoner of war camps, firing ranges, ordnance factories, and research and development labs. Route 66 was transformed from the Mother Road to a military highway. The road would be ridden hard, but it would not be improved until the war was won four years later.

Alma Mildre owned a service station on Route 66, just as the country was gearing up for World War II. "The only thing that brought us out of the Depression was when they started the war and put everybody to work," she said. "When we was first in the service station business, we'd look out and say, 'Can you see a car coming in any direction?' And we'd look and maybe it would be

Route 66 was transformed from the Mother Road to a military highway. The road would be ridden hard, but it would not be improved until the war was won four years later.

nine thirty at night or eleven or twelve and if there was a car coming that might need help, we'd stay to service it, and if there wasn't a car in sight in any direction, we'd turn out the lights and go on home. Now you just look down the road and it would be impossible to close."

Herb Tiffany had a paving business along Route 66 in the 1920s. "Starting out, times were tough. Dad made his own butter, and the highest-paid worker on the payroll was a mule at eight bucks a day," recalled his son, Herb Tiffany, Jr.

"In aviation, Dad was a pioneer of sorts. He flew for the U.S. Army Air Corps during World War I teaching pilots the finer points of take-offs and landings, you know, touch-and-gos. When he got out, he became the 222nd licensed pilot in the country. He enlisted for active duty in World War II, but they said he was too old to fly. During the war, though, there were a bunch of airfields for training pilots up and down 66—the Army built them out here in the desert because the weather was ideal." The airborne patriot who spent the first World War in the sky, training fighter pilots, spent the second one asphalting the

ground they depended on most—their runways.

"In those days, Tiffany Construction could build a paving camp—sleeping and dining quarters for a crew of 15—for $650. I still have the company's ledgers to prove it. It only took 65 bucks to stock it with enough food to feed the guys for a month. The crews could disassemble a camp, load it onto trucks, and haul it off to the next paving job." Tiffany Construction paved asphalt runways at Gila Bend, Luke Field, Ajo and Whitman airfields, to name just a few.

After the war, eight million Americans relocated to the West and nearly half settled in California. A great many of them traveled down Route 66 to get there. Larger than the migration during the Great Depression, the postwar population resettlement along Route 66 and the other interstate highways was changing the fabric of the nation forever. The only problem was that the old two-lane interstate highways were torn up from the war and too narrow for all the traffic. After the war, the country began to demand better, wider roads—and more asphalt.

CHAPTER ELEVEN

Paving the Way to Victory

"Undoubtedly, the ability to build pavements quickly where they were needed was a major factor in bringing the war to a successful conclusion."

WHEN THE FIRST Japanese torpedo bombers swept in over Oahu to attack Pearl Harbor, they did not catch a nation totally unprepared for war. Overseas developments during the turbulent decade of the 1930s had pushed our once-isolationist nation into looking outward again. The onset of war in Europe, starting with the Nazi invasion of Poland in September, 1939, provided the spark. By December 7, 1941, the Day of Infamy, the United States already had a two-year head start in building the mightiest war machine on the planet.

172

THE GREAT MOBILIZATION

If America was to have any chance at all in the upcoming global war, it had to build a large army and an ancillary air armada from what had been a relatively minor force. It also had to revive a dormant munitions industry to arm them. The foundation of this growth was the construction of needed bases and support facilities.

For the United States Army and its Air Corps, this meant camps, cantonments, depots, arsenals, and airfields. The U.S. Navy required bases, ports, and airfields of its own. There were munitions plants, transportation facilities, hospitals, and headquarters to be built. As the army grew from 200,000 soldiers in the 1930s to over 1.6 million by 1941, construction expenditures grew to $15.6 billion (more than $180 billion in 2005 terms).

Seabees paved miles of runways and roads with asphalt in both the Pacific and Atlantic theaters of World War II. Here, the 53rd Naval Construction Battalion blacktops a road on Guam in 1945.

Asphalt was an important ingredient in construction for mobilization. The army's experience with it dated back to 1876, when a group of army engineers began to study the use of asphalt as a paving material. Their recommendations led to the paving of Pennsylvania Avenue in Washington, D.C., with Trinidad lake asphalt. In the ensuing years, acceptance of asphalt paving by the military would parallel that in the civilian sector.

The Army Corps of Engineers found asphalt an ideal paving material for its facilities. It was versatile, weather- and wear-resistant, and above all, quicker to construct and cheaper than concrete.

Facilities required paved streets and access roads, parking lots, and hardstands for parking planes. The Quartermaster Corps and its domestic construction successor, the Army Corps of Engineers, found asphalt an ideal paving material for its facilities. It was versatile, weather- and wear-resistant, and above all, quicker to construct and cheaper than concrete. Design methods and construction techniques developed by civilian industry and state highway departments could be used in military facilities. After all, a road built to support a three-axle munitions truck was no different from one to support a produce truck.

Although some munitions plants required special pavements that were perfectly level and smooth, those were the exception. Most military asphalt paving within the United States was routine work done by civilian contractors to civilian standards.

However, in the months before Pearl Harbor, the real story of military asphalt was unfolding. It was not primarily a tale of roads, parking lots, and hardstands, but one of runways, taxiways, and aprons or tarmacs, and the airplanes that needed them. It begins with one plane in particular.

THE BREAKTHROUGH BOMBER BREAKS THROUGH

The Douglas XB-19 was the Army Air Corps' first experimental model of a new and revolutionary class of war machine, the very heavy bomber (VHB). This flying behemoth was 132 feet long with a wingspan of 212 feet. The top of its tail rose four stories above the ground. It could carry a 16-man crew and 18 tons of bombs a distance of 2000 miles. It came armed to the teeth, its fuselage bristling with two cannons and 11 machine guns.

On May 6, 1941, the doors of the Douglas Aircraft Company hangar in Santa Monica were opened and the silver bomber rolled out. Those seeing it for the first time must have reacted with awe; it was the largest American plane that had ever been built. It made its way onto the parking apron—and promptly broke through the pavement, its eight-foot-tall wheels sinking a foot below the surface.

The ground crew managed to tow the plane onto an asphalt runway, where its single-tire main landing gear would crack and rut the pavement during subsequent taxi tests.

It took seven weeks and the construction of a concrete runway before the XB-19 was able to take off on its first flight to nearby March Field, where it was turned over to the army for evaluation. An army officer observing the arrival of the bomber noted depressions as deep as one inch and large cracks in the runway as the plane decelerated. All of this was caused by an empty plane weighing a mere 92,000 pounds. There was no telling what damage a fully loaded 160,000-pound plane would do.

The air corps eventually rejected the XB-19 on its own merits—it was underpowered, far too slow, and highly vulnerable despite the host of protruding guns. The prototypes were destined to serve as flying morale boosters on the home front before ending up on the scrap heap. The runway designers must have breathed a sigh of relief when they were granted a respite from having to deal with the damaging effects of the XB-19. But there were other superbombers on the way, and the airfields had to be made ready to handle them.

Above:
The XB-19 was the first very heavy bomber and America's largest plane when it debuted in 1941. Inset: The huge tire for the XB-19 was eight feet in diameter.

SOMETHING BETTER THAN SOD

It seemed that the runway designers and builders hadn't kept up with the advances being made in aircraft technology. The air corps entered the decade of the 1930s flying flimsy, fabric-covered biplanes like the ones King Kong swatted from his perch atop the Empire State Building. By decade's end, the Corps planned a fleet of all-metal streamlined warplanes that were faster, more powerful, and, of course, much heavier.

Not that there were too many of these metal machines. In 1939, as the European conflict was beginning, the U.S. only had 19 heavy bombers. With only 20,000 men and 800 mostly obsolete planes, even one of its top commanders, Major General Frank Andrews, deemed the Air Corps a "fifth rate air force." The army brass tended to underestimate the importance of airpower and allocated relatively few resources to its development.

Given the scant attention paid the air corps itself, it was small wonder that runways were overlooked. As late as 1939, runway designers were told that the airplanes they were designing for were so light that sod surfaces would be adequate for advance bases. For heavy bombers based in the rear areas, designers believed that they could rely on existing highway pavement methods. This was because the standard aircraft wheel load was 12,500 pounds, about the same as that of a heavy commercial truck. With such low design loads, some air corps officers believed that paved runways were a time-consuming and expensive luxury, and even questioned the need for purchasing paving equipment.

When the Army Corps of Engineers took over the task of building airfields in 1940, it knew that it had a lot of catching up to do. By the time of the XB-19 rollout fiasco, the engineers recognized that the standard 12,500-pound wheel load was a grossly insufficient standard for runway design. After all, heavy bombers with wheel loads of 37,500 pounds were already in production. A new generation of very heavy bombers was currently on the drawing boards. Although not as heavy as the Douglas plane, the Boeing B-29 would have twin-wheel loads in the 60,000-pound range; the six-engine XB-36 intercontinental bomber would be even heavier.

The military brass knew very heavy bombers would be needed to win the war in the Pacific, because an invasion of Japan would most likely be necessary. Before that invasion could occur, strategic Japanese mainland military and industrial installations would have to be eliminated. This would require massive loads of explosives and incendiaries delivered over a great distance. Only very heavy bombers would have the bomb capacity and range to accomplish the

Some air corps officers believed that paved runways were a time-consuming and expensive luxury, and even questioned the need for purchasing paving equipment.

mission, and those bombers, in turn, needed proper runways. It was as simple as that.

THE GREAT RUNWAY DEBATE

The Corps of Engineers embarked on a massive research and development effort to establish design standards for the new runways. Like the Manhattan Project, it would be conducted in laboratories and test sites all over the country by dedicated teams of engineers from the military, civilian industry, and academia. It would take years of frantic and intense effort to complete. Unlike the Manhattan Project, there was little secrecy involved (and no Nobel prizewinners).

The first step was to publish some standards—any standards—for runways and taxiways. In January 1941, the Corps of Engineers, under Colonel William McAlpine, published "Design of Airport Runways" with hastily compiled materials from the Civil Aeronautics Authority, Public Roads Administration, Portland Cement Association, and the Asphalt Institute. The manual featured sections on grading, drainage, and runway layout with formulas for flexible and rigid pavements. However, the manual was meant only "to serve as a general guide and not a source of specific instruction."

In the meantime, the air corps demanded that only rigid pavements—those constructed of Portland cement concrete—be used for runways and taxiways. They claimed that only concrete, even on weak subgrades, would stand up against heavy wheel loads of the four-engine bombers. Furthermore, they believed that concrete would have high skid resistance, greater ease of maintenance, and, since it was light in color, better visibility at night.

The engineers took exception. They didn't want anyone telling them how to build airfields. Their attitude was, "Tell us what you want it to do and let us figure it out"—an admonition of engineers to their clients since the time of the pyramids.

The engineers weren't convinced that concrete pavements were the answer. Concrete suffered from long cure times, specialized equipment requirements, and high costs. It would be particularly difficult to use in active theaters of operations, where runways would be needed immediately. It had no better skid resistance; nor was its light color necessarily an advantage for night operations.

Asphalt was cheaper and easier to apply. A runway could be built during the night and used the next morning. In addition to hot-mix asphalt, a wide range of surface treatments, seal coats, and dust palliatives could be applied to asphalt pavements to meet varied theater requirements.

Highway experience showed that the quality of the subgrade was key. Application of asphalt courses on a foundation of proper thickness, plus quality design, compaction, and drainage could produce flexible asphalt pavements that would withstand the heaviest bombers and could be maintained more cheaply than concrete.

The Engineers contended that the choice of paving materials and construction methods should be dictated by ground conditions. In the U.S., where both were readily available, concrete and asphalt would be preferred options. To develop workable design standards for both materials, a competition of sorts should ensue. That was the American way.

The controversy made it all the way up to the Assistant Chief of Staff for Logistics, who ruled in favor of the Engineers. He directed the airmen to come up with a list of functional requirements. The Engineers would then figure out how to build the runways to meet the specification. The spec was published in August 1941 and provided a tough set of requirements for runways.

- Sufficient strength to carry wheel loads up to 100,000 pounds as well as stress loads of 500 psi (pounds per square inch) under impact
- Proper waterproofing and drainage to prevent excess water infiltrating the subgrade
- Features to protect aircraft—low rolling friction, high skid resistance in wet weather, low crowns to prevent ground loops, low levels of dust and debris that might be kicked up by wheels or propeller wash
- High visibility at night
- No maintenance except for bomb-damage repairs.

Colonel McAlpine instituted a broad research program using both military and civilian experts. Separate investigative programs were formed to address the various aspects of runway design. One group, the Corps of Engineers' Waterways Experiment Station (WES) in Vicksburg, Mississippi, had already been studying soil stabilization, runoff and drainage, and flexible pavement design issues since January 1941. Another team explored the role of surface runoff and drainage, which was critical for wide pavement surfaces on level grades.

Traffic tests of rigid pavement were performed at two of the Corps of Engineers' Ohio River Division facilities, the concrete and soil mechanics laboratories. Concurrently, tests were performed on asphalt pavements at

Langley Field was a site where asphalt was tested for air force runways and taxiways. Here, an A-20 bomber is being serviced at Langley in July 1942.

Langley Field, Virginia, in conjunction with the state highway department, the Asphalt Institute, and the WES.

Other investigations looked at the aircraft themselves. The pounding dynamic loads on pavement that were caused by the impact of landing were initially a concern. However, experience with the XB-19 indicated that cracks on the pavement appeared not on impact but when the bomber slowed down. Subsequent controlled tests with medium bombers showed that impact was not as important as originally thought because of ground effect, or wing lift. While the dynamic effects of takeoffs and landings were not a factor, the numbers of takeoffs and landings were.

Investigators also demonstrated that high repetition of smaller loads was just as damaging to runways as occasional heavy loads. Taxiways, where airplane wheels tended to travel over the same narrow strips, were especially vulnerable.

The size of the loading area was important. Using a single wheel on each side of a landing gear (like the XB-19's eight-foot monstrosity) would concentrate the weight of an airplane over a relatively small area. The proposed B-29 would be designed with two wheels on each side of its landing gear, providing a larger footprint and distributing the weight over a larger area.

Although technical progress was being made, political problems mounted. Shortly after Pearl Harbor, the air corps pressed the Corps of Engineers to suspend its testing program. A real live shooting war was in progress, one that could not wait for an investigation of different pavements. There was enough data on concrete behavior already available to create usable standards, and concrete pavements were what the airmen wanted. Once more, the top brass stepped in. This time the A-4, the Assistant Chief of Air Staff for Matériel, put an end to the debate and upheld the Corps of Engineers' decision to study asphalt runways. The most vocal pro-concrete officer was removed, and the airmen and the engineers declared a cease-fire.

CRASH PROGRAM

In the meantime, the situation was getting desperate. Soon after Pearl Harbor, all available B-17 Flying Fortresses in the west were ordered to the Pacific Theater to reinforce American forces besieged by the Japanese. A mass movement of these bombers originated from West Coast bases. The B-17s cracked, rutted, and

Porter developed the California Bearing Ratio (CBR) method for assessing the strengths of base and subbase materials.

A CBR apparatus tests soil strength at a paving site. The revolutionary CBR method was used to show that asphalt could withstand the stress of aircraft landings, takeoffs, and taxiing.

otherwise wrecked both asphalt and concrete surfaces, despite the fact that they were only half the weight of the proposed very heavy bombers. The pressure was on for the engineers to develop serviceable runways for the superbombers by the time they would come online in 1944.

In December 1941, Lieutenant Colonel James Stratton replaced Colonel McAlpine as head of the army's engineering branch. The 43-year-old West Point graduate already had runway experience behind him, having built a bomber-capable asphalt runway on a caliche base in West Texas, much to the surprise of skeptical air corps personnel. The new chief immediately took an interest in the flexible pavement program.

The Langley Field tests were finally getting under way, and the results were dismal. Although the Asphalt Institute claimed that thick bituminous surfaces could provide strength through beam action, pavements designed for 60,000-pound wheel loads were failing at a third of the design load.

At the recommendation of his staff, Stratton sent for O. James Porter, an assistant research engineer at the California Division of Highways. As a junior engineer, Porter had studied pavement failures and came to the conclusion that thin asphalt surfaces over compacted bases and subbases provided the most durable roads. He developed the California Bearing Ratio (CBR) method for assessing the strengths of base and subbase materials. Using this method, he created design curves relating pavement thickness required to support a design load to base and subbase quality.

CBR had only been applied to roads, but Stratton's soil advisor, Professor Arthur Casagrande of Harvard, felt that it could be extended to runways through additional research. Three consultants from different backgrounds—Casagrande, Porter, and soils specialist Thomas Middlebrooks—were able to independently extrapolate the existing pavement design curves to wheel loads up to 70,000 pounds. Within a week, the Corps of Engineers adopted the CBR method as the design basis for runway pavements.

In March 1942, Stratton ordered a series of tests conducted at a test track at the WES and at five different army airfields to validate CBR. The Vicksburg test section allowed WES researchers to test a variety of conditions, including different base-course materials. The testers used earth-moving equipment with tire sizes and contact areas similar to those of airplanes. Engineers towed Tournapull scrapers loaded with sand and 155-mm shell cases to simulate repetitive wheel loads of 5000 to 50,000 pounds. Drivers pulled the load in lanes until the pavement failed or a specific number of passes was completed. Then they made visu-

A Tournapull scraper simulated the wear and tear of heavy airplanes. These tests proved that the CBR system was valid in predicting asphalt paving's strength.

al tests and measurements and compared the results to lab tests. The Tournapull testing method was adopted for the other sites, where tests were done on asphalt pavements of differing thicknesses and subsoils (sandy clay, lean black clay, Fargo clay, porous subgrade subject to frost, and adobe/sandy loam).

By April, the tests would confirm the applicability of the CBR-based design curves. Now it was time to disseminate the information to the various engineering divisions involved in airfield construction. The CBR method was met with widespread doubt. The U.S. Army Air Force (as the air corps was now called) and other pro-concrete interests were skeptical at best; the navy's Bureau of Yards and Docks was downright opposed. State highway officials who knew of the tests were also negative. Even the engineering divisions were reluctant. Despite the early opposition, use of the CBR method would continue full tilt at U.S. airbases.

McFADDEN'S TRAVELING ROAD SHOW

As the flexible-pavement research effort continued at full speed, Stratton reorganized his staff, establishing an Airfields Group under Gayle McFadden, who had presided over the prewar construction of LaGuardia and National airports. By the fall of 1942, the flexible-pavement research program took on a sudden urgency.

On September 21, 1942, the prototype Boeing Model 345 very heavy bomber, to be designated the XB-29, lifted off the runway in Seattle on its maiden flight. The air force was pinning America's hopes on this "Superfortress," ordering an initial production run of 250. The army issued directives for more than 100 major airfield projects to be completed as quickly as possible. Most of the newly ordered runways would be needed for bomber-crew training when the first production models entered service a year later.

It didn't help that several of the largest bomber training facilities were to be located in the northern Great Plains and Pacific Northwest, where airfield construction continued year round under the worst winter weather conditions. To lend assistance, provide advice, and prevent construction blunders, McFadden organized a group of consultants to travel from site to site.

Civilian O. James Porter, the originator of the CBR method, would spend so much time on the road (his wife claimed that he was gone 300 out of 365 days a year) that the army brass suggested that he receive a commission as a full colonel as a token of their gratitude. The idea was immediately quashed by Stratton and the chief of the engineers, who feared that once he was in the army, he could be transferred anywhere but the place where he was needed.

THE FLEXIBLE PAVEMENT LAB IS BORN

Opposite:
The Marshall
Method of testing hot-mix asphalt became an industry standard. At upper right is a stability and flow tester, plus the tall Marshall hammer used for compacting a sample. At upper left, Marshall testing is perfomed in a lab. The lower picture shows the hot-mix asphalt plant at WES.

While McFadden's group was crisscrossing the country by plane, train, or automobile, the Waterways Experiment Station was becoming the preeminent facility for flexible-pavement research. It had already become the focal point of CBR research and testing as well as airfield drainage design. By the spring of 1943, the Corps of Engineers had formally established the Flexible Pavement Branch at WES under Willard J. Turnbull. A new fully equipped Flexible Pavement Laboratory (FPL) building was dedicated that April.

By the summer, the FPL had staffed up to 25 researchers and one cat. The team included Bruce Marshall, who had recently invented a machine for measuring asphalt mix quality. Up to then, the rational test methods used in designing mixes were the Hubbard-Field test (developed in the early 1920s, and only useful for sand mixtures) and the Hveem test (developed in the 1930s, and adopted by a limited number of states because of its time-consuming and expensive methods). In the absence of a reliable test, the proper asphalt mix was a matter of guesswork. The quantity of asphalt in the mix was important: too much, and the pavement would be soft and deform under traffic; too little, and the pavement would be brittle and crack.

Marshall had developed a practical method for evaluating the bituminous mix when he was at the Mississippi Highway Department but had been unable to convince the organization to adopt the device as the standard. Undaunted,

Engineers who developed the specifications and protocols watch pavement testing. Standing on the tire, at left is Arthur Casagrande, with Thomas Middlebrooks, James L. Land, O. James Porter. Front row: Colonel Henry C. Wolfe, Harald M. Westergaurd, Philip C. Rutledge.

the paving specialist showed up at FPL's doorstep with his device and was hired on the spot. Testing by FPL, some of which extended beyond the war, proved that not only was his method superior to other methods, but the equipment had the advantage of light weight and portability. The Marshall method become an industry standard.

After establishing CBR-based standards for asphalt runways, the Flexible Pavement Branch and the WES would go on to tackle the problem of "expedient" runways—surfaces that could be constructed for immediate use in war zones. In many cases, runways would be built of indigenous materials or materials that were easily transported to the area. This ruled out both rigid and flexible pavements. WES helped develop and tested a number of systems, including pierced-steel planks and asphalt-coated fabric. At the same time, Stratton's group started to look at indigenous materials and construction techniques. Materials included hand-set Telford stone, waterbound macadam, sand-clay and sand-asphalt bases. The data gathered would be crucial in the China-Burma and Pacific Theaters of Operations, where army Aviation Engineers and navy Seabees were faced with a lack of raw materials for pavements.

By the end of 1943, the bulk of the initial runway-development effort was finished. The engineers had overseen the construction of over 1100 runways, many of them asphalt, in the continental United States. For a job well done, Colonel Stratton would be rewarded with a position as a logistics officer on General Eisenhower's staff.

Thanks to his work and to the efforts of Casagrande, Porter, McFadden, Turnbull, the WES, and others, asphalt runways at B-29 bases across the nation had been ready when the first training groups were formed in September 1943. Within a year, more than 1000 B-29s would be stationed at advance bases in the Pacific.

THE SEABEES ON THE ROAD TO VICTORY

While the Army Corps of Engineers was charged with building bases and facilities domestically, the job of establishing advanced bases was primarily assigned to the navy. In the areas where they were building—some of them active war zones—the navy had a unique problem. Civilian contractors could be used to

The navy's Seabees helped fight for enemy territory, then built the pavement for necessary roads and airfields. They're known for their "can do" spirit, as touted on this poster.

build bases domestically, but in the forward areas they would be in danger and they had no protections under the Geneva Convention. If captured, they could be treated as guerillas—executed instead of being taken prisoner.

To meet its construction needs, the navy formed Naval Mobile Construction Battalions, the NMCBs, shortly after the attack on Pearl Harbor. Nicknamed after the acronym for Construction Battalion, the "Seabees" landed their men and equipment in the advanced areas, built facilities sometimes in the face of enemy fire, and fought for the land they needed to build on if they had to.

Recruiters built the Seabees from the ranks of grizzled veterans—of the pre-war civilian construction industry. The navy felt that it would be easier to train experienced tradesmen to fight than to train raw recruits to build. Because of this, the Seabees tended to be older, with an average age of 37. A few men in their 60s were able to sneak through to join the ranks. They would soon make a name for themselves in the Pacific and live up to their motto *Construimus, Batuimus*—We build, We fight.

While the army domestic building program was reaching its peak in 1942, the Seabees were just getting under way. Navy planners had identified five "roads to victory" that would bring American forces from the continental United States to the enemy homelands. In the European Theater of Operations, the Mid-Atlantic Road led to Germany in a circuitous path through the Caribbean, the Mediterranean, and Italy, while the North Atlantic Road traversed Newfoundland to England and France. In the Pacific, Japan would be attacked from the South Pacific Road through the Solomons and Philippines, the Mid-Pacific Road from Hawaii to the Marianas, and the North Pacific Road from Alaska and the Aleutians.

THE NORTH ATLANTIC ROAD: AIR BASES ON THE SNOWBALL ROUTE

On the Atlantic roads, close to home, asphalt materials were generally available, and it was safe enough to use nearby contractors and suppliers or even bring in crushers, blending and mixing plants, delivery trucks, pavers, and steamrollers. In the North Atlantic, the Germans made their presence known, but the real foes were the extreme site conditions faced by the construction battalions.

On the North Atlantic Road, the Seabees began construction on a series of bases in Newfoundland, Greenland, and Iceland along the so-called Snowball Route. These bases would be used as way stations for aircraft being ferried to the

This Greenland airfield, on the Snowball Route, was one of the most difficult to build. A steel-mat runway was constructed on the frozen earth-and-stone surface left from a glacier.

European Theater. They were extremely important; despite the harsh weather encountered by aircrews, the Snowball Route presented the safest way to get desperately needed aircraft to England in the early days of the war. Transporting aircraft by surface ship was both time-consuming and dangerous in the seas ruled by the German U-boat wolf packs.

Along the route, Seabee construction crews were forced to build on miserable terrain covered with tundra and rife with boulders and ledges. They had to deal with permafrost, wretched cold, and violent winds.

In Iceland, construction at Meeks Field, a large airfield for bombers and cargo planes, commenced in July of 1942. Work on a separate base for fighters at nearby Patterson Field wouldn't begin for another four months; by then, it was a desperate attempt to get one runway paved before heavy rains and snows caused the fill to turn into a bog.

The most difficult obstacle faced by the Seabees was freezing of the top layers of the compacted runway fill. When hot asphalt was applied on the frozen surface, the frost melted and formed pools of mud beneath the asphalt, thereby weakening it. This problem was solved by laying a porous-mix base course that allowed the water to escape as steam.

At Patterson, asphalt paving was the bottleneck. A minimum of 1500 square yards per day of asphalt paving was required to meet the schedule. By the end of November, a 3500-by-100-foot pavement had been completed. However, heavy use and hard weather caused equipment to fail constantly.

By February 1943, both the asphalt plant and the rock crusher had broken

down. A new asphalt plant that had been ordered the previous December arrived in May to share the load. By then, the fill at Patterson had dried out sufficiently to allow paving and grading to proceed again. When paving operations were completed in July 1943, nearly two million square yards of base course and topping had been applied at Meeks. By November, a million square yards had been placed at Patterson.

When full operations were established in 1943, over 100 planes a day could pass through Patterson and Meeks Fields when the weather cooperated. They included the B-17 and B-24 bombers; the P-38, P-47, and P-51 fighters that would one day command the skies over Germany; and the C-47 and C-54 transports that would support the invasion of Europe.

SEEKING THE SHORTCUT TO JAPAN

The sheer size of the Pacific Theater of Operations required that the U.S. adopt a multipronged strategy to reach Japan. On the Asian mainland, the army built its own "road," a series of bases in the CBI (China Burma India) Theater to threaten Japan from the west. From the east, approaches would be made from the South, Mid- and North Pacific roads. As time passed, both the Asian mainland and the North Pacific Road looked less favorable for basing attacks on Japan. Allied forces eventually planned to invade the Japanese home islands from the South or Mid-Pacific Roads. Hopefully, one would turn out to be the shortcut to Tokyo and victory.

The Pacific roads are where the Seabees established their legend as builders and fighters. To build the bases, the U.S. had to pursue the arduous strategy of island hopping. Real estate would be acquired one island at a time, in many cases by forcibly evicting its previous occupants. Once a foothold had been established, a base could be built from which to launch an assault on the next island.

The Seabees closely followed the assault forces step by step across the Pacific, each step bringing the Allies closer to Japan. In all of the great land battles of the Pacific Theater—Guadalcanal, the Solomons, New Guinea, the Hebrides, the Philippines, the Marshalls and Marianas, Okinawa, and more—the Seabees were there, often landing with the first waves of marines.

The Seabees repaired captured Japanese bases and airstrips for Allied use. Sometimes runways had not been 100 percent captured—forcing the Seabees to work under fire to patch pavement and clear debris to open runways for waiting American planes. The battalions also built new runways, working 24

Opposite:
The Pacific Theater involved long distances from Allied bases to the many Japanese-controlled islands. As the Allies took additional islands, the Seabees repaired existing airbases and built new ones for the push to the ultimate objective—Japan.

hours a day to carve them out of thick jungles, coral atolls, and volcanic rock.

Seizing existing Japanese airstrips reduced Japanese airpower in the area. Airstrips that were built or repaired could be used to establish air superiority by Allied aircraft. As the war progressed and the island hopping got closer to Japan, some of the captured or newly established airstrips would be dedicated to another purpose. In 1944, they would become the homes of the B-29 very heavy bombers that would attack the Japanese mainland.

It would take over three long and grueling years, but their innate ingenuity and can-do attitude carried the Seabees across the Pacific. They had to fight impossible schedules, lack of materials, broken equipment, and the Japanese, but they did whatever it took to do the job. This often meant working 24 hours a day, repairing equipment with tape and baling wire, inventing new equipment on the spot, or "appropriating" needed supplies on midnight raids. And they weren't afraid to fight, either. Seabees would win 33 Silver Stars and five Navy Crosses. Three hundred of them died in combat and 500 in construction accidents.

PAVING THE PACIFIC ROADS

The Seabees and their army counterparts, the Aviation Engineers, built 111 airfields in the Pacific during World War II. Of these, only a handful were paved with asphalt, and almost none used concrete. It seems that the Great Runway Debate going on stateside—with millions of dollars spent and careers and reputations made or broken—was largely irrelevant to the Seabees and Aviation Engineers.

As fast and as easy as asphalt pavements were to build, the materials and the equipment had to be shipped in with the construction crews. Aggregate was available, but drums of liquid asphalt and batch plants had to arrive with the landing crew. If asphalt wasn't available, the war effort wasn't going to wait.

The Seabees logo shows a sailor-hatted "sea bee" equipped to fight or build, as needed.

So they used whatever they could get their hands on—bare soil, coral, pierced planks, gravel, even volcanic cinders and other unspecified "native materials." These were particularly important when constructing expedient airstrips. Once the expedient strips were established, the crews could go to work on larger and more permanent runways. When available, asphalt was used to surface coral and concrete, adding its superb water resistance, slip-proofing, and ability to keep dust and debris down.

Proximity to home or to heavily populated Allied territories determined the availability of asphalt. In Hawaii and the outlying islands, 23 asphalt airstrips were built, along with parking areas and taxiways. Asphalt roads

Throughout the Pacific Theater, as islands were captured, the Seabees created airfields. Top left: Bulldozers prepare the land on Eniwetok Island. Above: Rocks are crushed for asphalt to build roads and runways. Left: Asphalt is laid on the island of Guam.

and lots were built to support the many military facilities. Australia and New Zealand had plenty of equipment and materials for asphalt paving. So did the larger Pacific islands like Guam.

Farther out, asphalt was harder to come by, but when it came time to surface the runways for very heavy bombers, the descendants of the XB-19, asphalt paving played a very important role. The first B-29 Superfortresses arrived in the Pacific Theater in 1944, ready to disrupt Japanese shipping and strike strategic installations in the area and on the Japanese islands.

Here are the stories of three blacktopped airfields in the Pacific. Two of them were in the Northern Marianas Islands—within B-29 striking distance of the mainland; one was a safe haven halfway there. One of them would prove to be the shortcut to Japan.

SAIPAN

To provide a superior landing and takeoff surface, asphalt was shipped in— ignoring Washington naysayers, who claimed that "blacktop couldn't be used in the Marianas."

The island of Saipan in the Northern Marianas was the key to invading Japan. The 15-mile-long island lay a mere 1500 miles from Tokyo. From its airbases, the B-29s were within comfortable striking distance of the home islands. On June 15, 1944, a force of 20,000 U.S. Marines landed on the island and ultimately took it after a fierce battle. Within 24 hours, U.S. Army Aviation Engineers landed and began to fix the captured Aslito airstrip. Despite fierce Japanese counterattacks, the airstrip was operational within days.

Soon work began on building a new bomber strip at the renamed Isley Field. For fill, two mountains of coral were virtually leveled and crushed by bulldozers, and the material was brought to the field by 100 four-ton trucks. To provide a superior landing and takeoff surface, asphalt was shipped in—ignoring Washington naysayers, who claimed that "blacktop couldn't be used in the Marianas." The army construction crews brought in hard asphalt in drums and, in a manner similar to their Seabee counterparts, put their ingenuity to work. They modified a sugar boiler abandoned by the Japanese forces, adding a smokestack of welded oil drums to create an asphalt-melting plant. The jury-rigged contraption ended up supplying 700 tons of liquid asphalt a day for a surface treatment over a coral base.

When it was finished, the 8500-foot-long Isley No. 1 airstrip was the longest runway in the Pacific. In November, the first of the B-29s arrived. On November 24, they hit Tokyo. The tide of the war had truly turned for the Allies.

THE TRANSFORMATION OF IWO JIMA

In early 1945, the Japanese air base at Iwo Jima posed a direct threat to the U.S. bases in the Marianas. The Japanese had built three airfields on the small volcanic island; fighters based there could attack Saipan-based B-29s on their way to and from Japan. Capturing Iwo Jima would not only eliminate the threat to the Superfortresses, it would provide a base closer to Japan from which to launch escort fighters for the bombing missions. At the same time, it would provide a safe haven for crippled bombers that couldn't make it back to Saipan or Tinian. In mid-February, the U.S. Marines launched an all-out amphibious assault on the island.

The month-long battle for Iwo Jima was one of the bloodiest of the war, the struggle captured in the famous photograph of the raising of the flag. Seabees went ashore with the marines to begin work on the airstrips as soon as they were captured. Intense enemy opposition made construction difficult, but the Seabees worked to get the strips into minimum service condition. This happened none too soon; on March 9, a crippled B-29 was forced to make an emergency landing.

After the Battle of Iwo Jima, the Seabees were still under enemy fire as they prepared airfields such as this one. The fields were especially important for aircraft that were crippled in the assault on Japan or were diverted from their home fields by bad weather.

The unluckiest pavement in the world. Even as runways on Iwo Jima were being completed by the Seabees, pavement was being intentionally destroyed in New Mexico. The first atomic explosion occurred at the Trinity Test Site on July 16, 1945, "instantly vaporizing the tower and turning the surrounding asphalt into green sand."

One of the unique problems facing the combat pavers on Iwo Jima was the fact that it was an active volcano. The island had risen from the sea in recent memory, and the ground was laden with pockets of steam. While the volcanic ash used for base was easier than coral to compact, it also eroded quickly when wet. In addition, the ash base had to be completely dry when asphalt was applied. Because of this, heavy rains held up paving.

Subgrade problems due to steam condensation and wet conditions caused further difficulties. At one point, this led to the removal of 1500 feet of crushed stone and subgrade. Despite the setbacks, the first superbomber runway, 8500 feet in length, was completed by July 1945. On July 7, after a strike on the Japanese mainland, 102 B-29s landed. Although it was not a main staging field for bomber strikes on Japan, Iwo Jima would go on to serve as a safe haven for crippled aircraft and a diversion point in case of bad weather. Over 2400 B-29s carrying more than 25,000 airmen would land safely on the two bomber strips.

TINIAN: LAUNCH PAD TO VICTORY

A month after its capture, Saipan became the staging area for the invasion of the nearby island of Tinian. The marines had barely established a foothold when the Seabees, in their uplifted baseball caps, came ashore with bulldozers and dump trucks. Among their equipment was an asphalt plant that was used to pave not only the four captured runways but also two new ones on the north and west sides of the island. To level the runways, the Seabees removed nearly four million cubic yards of hard coral and filled close to eight million cubic yards. Atop the base of packed coral fill, they added a 2.5-inch-thick layer of asphalt, which they rolled down to two inches.

Within a few months, six 8500-foot bomber-wide (400-foot-plus) runways, 11 miles of taxiways, and parking for more than 450 bombers had been finished, and Tinian became the largest airport in the world. Over 15,000 Seabees had participated in the construction effort. In January 1945, the first Tinian-based

B-29 took off from North Field for a raid on Kobe, Japan. By August, the six runways were teeming with activity, as over 450 B-29s were based at North and West Fields.

The island-hopping strategy, the very heavy bombers, and the advanced bases and runways had all delivered on their promises. The B-29s had already gutted most of Japan's major cities and were flying virtually unopposed over the rest. Preparations had long been under way for Operation Downfall, the final invasion of the main islands. The first phase would be an amphibious assault of the southern island of Kyushu by close to a million men, scheduled for November. Despite the horrendous destruction brought by the bombers, the Japanese were expected to fight fanatically to the death for their homeland. Terrible casualties on both sides were anticipated, and yet seemed inevitable if the war was to be brought to conclusion.

Amid the hubbub of constant day and night missions, a new bomber group

Asphalt plants like this one turned Tinian Island into the world's largest airport.

The *Enola Gay*, a B-29, had just returned to Tinian after dropping an atomic bomb on Hiroshima when this picture was taken.

quietly arrived at Tinian in June 1945. The 509th Composite Group must have seemed an oddity to the other groups based on the island. It appeared to have no real missions, just training and recon flights. For those who were curious, the answer came in the early morning hours of Monday, August 6.

At 2:45 a.m., a B-29, named *Enola Gay* by the pilot in honor of his mother, took off from North Field. The previous afternoon, the bomber had sat on a special asphalt apron at the north end of the runway as a single 9000-pound bomb was painstakingly loaded into its bomb bay. With the bomb, a crew of 12, and 7000 gallons of fuel, *Enola Gay* weighed 65 tons—15,000 pounds overloaded for its high-altitude bombing mission. As it rolled down the runway, it was hard to believe that just four years earlier the XB-19 had sunk through the pavement in Santa Monica.

Six hours later, *Enola Gay* dropped the first atomic bomb on Hiroshima. A second Tinian-based Superfortress, Bockscar, dropped another atomic bomb on the city of Nagasaki three days later. There would be no invasion. Within six days, Japan surrendered.

DEMOBILIZATION

Victory brought about the demobilization of forces. Soldiers, sailors, and airmen alike went home. Of the more than 1400 airfields built for the military, over 500 were immediately declared surplus and transferred for use as civilian airfields. Some were closed down permanently.

Most of the combat pavers returned to civilian life—some working in the civilian construction industry, a few even becoming asphalt pavers. For those who remained in the military, there was much work to be done restoring the shattered infrastructure of allies and former enemies alike, and girding the U.S. for the Cold War.

The U.S. Air Force became a separate branch of the armed services in 1947, but it was dependent on the other services for construction engineering until 1965. That year, the Navy Seabees and Army Corps of Engineers were joined by Air Force Rapid Engineer Deployable Heavy Operational Repair Squadron, Engineer, or REDHORSE squadrons. For the remainder of the 20th century, in times of peace and conflict, the Aviation Engineers, Seabees, and REDHORSE squadrons were busy building roads, runways, and bases domestically and throughout the world.

The Flexible Pavement Branch of the WES continued to be the dominant mil-

The XB-36 Peacemaker intercontinental bomber, at right, made its first flight one year after World War II ended. Its size dwarfs the B-29 Superfortress, at left, one of the largest aircraft used in the war. The XB-36 would require even stronger runways to support its weight.

THE GREAT RUNWAY DEBATE, PART 2

"Congressman Hebert of Louisiana will show that the Armed Services have wasted $100 mil-lion this year by insisting on all cement runways instead of asphalt for military planes."
—Drew Pearson, "Washington Merry-Go-Round" columnist

THE EXIGENCIES of World War II temporarily shelved the Great Runway Debate in 1942, but it was far from over. From the Allied victory emerged a new foe—Communism—and America found itself rearming again. For this new "Cold War" a new generation of long-range superbombers would be introduced, each capable of delivering an atomic bomb to the Soviet Union from inside the continental United States.

August 1946 saw the rollout of the piston-powered B-36 Peacemaker, designed with this goal in mind. It weighed in at a whopping 300,000 pounds. A year later, the B-47 Stratojet made its first flight. Its swept wings and six turbojet engines made the Stratojet the harbinger of the future. Though classified as a medium bomber, its maximum gross weight of over 200,000 pounds made the B-47 much heavier than the B-29s that had won the war. It was the threat of these bombers and their terrible pay-loads that gave the Communists pause in their expansion plans.

In the late 1940s, asphalt runways were being built at some of the nation's busiest civilian airports—Newark, St. Louis's Lindbergh, and Balti-more-Washington's Friendship. At the same time, President Truman was being "hounded"

Congressman F. Edward Hebert

by asphalt and concrete proponents looking for a definitive ruling on which material would be used for commercial airports built totally or in part with government funds.

On the military side, several developments, political and technical, brought the asphalt-concrete debate to a head. The fledgling air force and the army now had equal stature under the newly formed Department of Defense. Disagree-ments over airstrip materials could no longer be settled within the confines of the army. Disputes would now be resolved at a much higher level—by the Secretary of Defense, or Congress itself.

By the 1950s, the pro-concrete air force had a heavy hitter on its side. General Curtis LeMay, a hero of the Pacific Theater in World War II, was now head of the Strategic Air Command. At LeMay's prodding, the air force instituted a policy requiring that the 1000-foot ends of runways and aprons—the areas most likely to experience fuel spillage either in landing or during refueling—be built of concrete, and that asphalt would be used only for the middle.

Asphalt industry protests led to hearings by the Sub-committee for Special Investi-gations of the House Armed Services Committee. The sub-

The B-47 Stratojet bomber posed a new challenge to airfields. On the ground, the plane's weight rests on the tandem landing gear at its centerline, with outrigger wheels for lateral stability only. Thus, over 200,000 pounds is concentrated on a thin strip of runway or taxiway.

committee concurred with the air force in its May 1954 report, and the air force restated its preference for concrete in August, although it "admitted to a lack of confirming evidence."

In December of 1955, the air force went a giant step further. It revised its criteria to specify concrete for all pavements on which aircraft were operated, parked, or serviced. This effectively prohibited the construction of asphalt runways. This ban helped to fuel the formation of a new hot-mix asphalt industry group, the National Bituminous Concrete Association (NBCA).

The NBCA took the fight to Congress and the Subcommittee for Special Investigations once more. The air force decision had drawn the atten-

tion and ire of Congressman F. Edward Hebert of Louisiana. A fiscal watchdog and chairman of the subcommittee, Hebert had previously taken the air force to task over cost overruns and excess profits in aircraft contracts.

Over a period of six days, the subcommittee took testimony from the service branches as well as industry groups, including the Portland Cement Association, the Asphalt Institute, and the NBCA. To augment its leadoff testimony, the NBCA prepared a 175-page presentation "written in non-technical language," entitled "Facts Show: Asphalt Can Take It." General LeMay himself never appeared to defend his arbitrary preference for concrete. Chairman Hebert instead trained his acerbic wit on an underling.

This time, the subcommittee recommended that additional testing be performed. The U.S. Air Force and the Army Corps of Engineers conducted a series of "proof" tests at Columbus Air Force Base in Mississippi between 1957 and 1958 to assess the capability of flexible pavements to handle normal operations of aircraft as big as the B-52. The tests confirmed that asphalt could take it, and when the investigative subcommittee issued its findings and recommendations in March 1959, the ban on asphalt was lifted.

The second and final chapter of the Great Runway Debate ended with the first legislative victory by the fledgling NBCA, which would later become the National Asphalt Pavement Association.

An oil-and-asphalt mixture is applied to an apron at an air base in Vietnam.

itary asphalt-research facility. The Vicksburg facility would continue its work in pavement design and research, transferring its design knowledge to the civil sector in time to support a postwar explosion in commercial air travel.

One WES engineer, Richard Ahlvin, developed a CBR design method that was the basis of flexible pavement design for airfields for 60 years after the end of the war. Another, John McRae, developed a gyratory compactor for laboratory use. With these and other tools, WES advanced technology for overlays that would allow runways built for the 120,000-pound B-29s to be used for the 500,00-pound B-52s of the 1950s and the 769,000-pound C-5 transports of the 1960s. The Flexible Pavement Branch later tackled a unique asphalt challenge that would have been inconceivable in 1940. Jet fuel dissolved asphalt and jet blasts eroded pavement surfaces. The air force tried to cope by first limiting the use of asphalt to the middle of runways, then banning asphalt entirely in favor of concrete *(see sidebar)*. WES soon developed blast- and fuel-resistant rubberized-tar-mix designs. Since that time a variety of specialized mixes have been developed for airport runways.

During the Vietnam War, the helicopter came into its own as a way of rapidly deploying troops and equipment to remote areas. However, its effectiveness was threatened by clouds of dust kicked up by rotors at landing zones and bases. Dust limited visibility, chewed up rotor blades, and clogged carburetors and engines. With the help of the Flexible Pavement Branch, the army developed peneprime, a high-penetration medium-cure asphalt surfacing, to reduce dust.

Today, the branch's successor, the Airfields & Pavements Branch, supports all the services—army, navy, air force, and marines—as well as the civil sector.

ASPHALT AND THE AGILE MILITARY

Sixty years after the end of World War II, military pavers are still building runways and roads all over the world, but in a different context. In this era of terrorism and small brush wars, of force projection and agile forces, asphalt remains an important material for the combat paver. While the material has evolved, and continues to do so—improving in strength, weight, and resilience—big changes have been made in the way it is transported and applied.

Combat pavers are concentrated in specialized battalions serving all branches of the armed forces. Engineers, soldiers, and equipment can be rapidly deployed at a moment's notice to a forward staging area halfway around the globe by military airlift or amphibious assault ship. The equipment is better, too. GPS (global positioning systems), CAD (computer-aided design), digital measurement systems, and electronic material-testing devices have now become part of the paver's tool kit. Even the graders, scrapers, pavers, and rollers, while appearing unchanged outwardly, feature composite parts and electronic controls.

What hasn't changed are the pavers themselves—the Seabees and the army and REDHORSE engineers. On the Baghdad Highway, in Tikrit and Kabul, the men and women of today still share the essential qualities of their World War II brethren: the tireless capacity to build, the willingness to fight, the ingenuity born only of the battlefield, and the spirit of "can do."

Army engineer operate a paving machine at Long heliport in Vietn in 1967. Helicopte rotors threw up quantities of dus which called for asphalt formulas

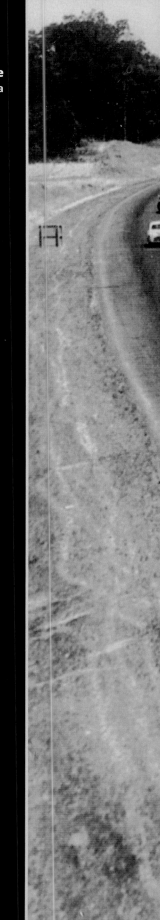

The **New Jersey Turnpike**
in the Woodbridge area

CHAPTER TWELVE

The First Asphalt Superhighways

"After careful consideration of all factors involving pavement, such as cost, durability and safety features, the Commissioners of the New Jersey Turnpike Authority have decided to use asphaltic concrete for the pavement of its 118 mile super-highway . . . the Turnpike will be superior to any road in existence."

"AFTER THE WAR THERE WAS A TRAFFIC JAM," recalled Bud Gunderson, who had grown up pumping gas at his father's filling station before World War II. As a returning veteran himself, Bud had a patriotic notion about travel: "One of the things every G.I. had fought for in World War II was to protect his rights as an American. And one of those rights was to be able to get in his car, turn the key and go anywhere he wanted to. There are no boundaries here; highways are not closed in the dark, or at state borders. As long as he had money for a car, he could go anywhere he wanted." And go they did. Topping off their gas tanks, veterans began visiting their buddies, moving to the suburbs, and taking family vacations.

Americans were ready to travel, but their roads weren't ready for them. Road-paving projects that had been envisaged before the war had been put on the back burner when Americans mobilized. The biggest automobile manufacturers stopped making cars and instead began making tanks, trucks, and warplanes. Gasoline was rationed and unnecessary travel was discouraged, even forbidden.

Rubber tires became a precious commodity, resulting in a drop in car and truck registrations. If a particular road survived the war years, it was because it hadn't been driven on much.

With the end of the war, gas was plentiful, new cars were being manufactured, and Americans began traveling. Massive maintenance and resurfacing programs were launched to address the damage inflicted over the past four years. The U.S. Army Corps of Engineers became involved in domestic paving programs.

FROM FIGHTING PANZERS TO RUNNING PLANTS AND PAVERS

The men who fought in World War II returned home and began waging peace. Having learned about engineering asphalt roads and operating asphalt-paving equipment, many of them decided to go into the asphalt-paving business. Those returning home to established family paving businesses were seen as wiser and more experienced in the ways of machines and materials. And it was a good thing because the roads they came home to were the exact same roads they had left behind—with the superimposition of a half decade of abuse and neglect.

The first half of the 1940s had been a time of "application and abandonment." Road-building projects not needed for the war effort were hastily wrapped up or abandoned. The only active paving projects were the ones needed for the war effort, and they received minimum amounts of asphalt. During the war, asphalt was used sparingly at best. "It was thin gruel or nothing at all," explained a Massachusetts highway engineer, Chan Rogers. However it was viewed, after a period of "five winters and four years" of either extreme use or complete neglect, the roads were in a disastrous state.

During the war, most paving contracts were conducted off the road, on military bases.

During the war, most paving contracts were conducted off the road, on military bases. Military contracts involving work on 1800 command installations and 2200 industrial facilities kept paving contractors busy. Once the G.I. was trained and armed, the paving action moved overseas.

A bright spot in the disruptive act of mobilization and wartime retooling was that the technology invented during the 1930s was being pushed into the future as it was tested under the extreme conditions of war. Machines used on rural paving jobs in the U.S.A. proved to be great for jobs overseas. Most notably, Barber-Greene's paving train, introduced in 1940 as Model 879A, emerged from the war as the most coveted paving machine in existence.

Construction battalions of both the U.S. Army and the U.S. Navy required that Barber-Greene machines be used in military operations worldwide.

Reflecting the military's desire for standardization, army construction battalions and the Seabees required that the Barber-Greene continuous paving plants, portable plants, and finishers be used in military operations worldwide. During the Depression years, Barber-Greene had an average order rate of just two travel plants a year. In four years, that number jumped to 200 travel plants and 500 finishers a year. Barber-Greene enjoyed a near monopoly during the war.

With the end of the war, a peacetime army of "engineers, technicians, supervisors, operators, mechanics" was ready to be employed in the rebuilding of the nation's roads. With patents in place and an abundance of surplus machines ready for use by thousands of demobilized servicemen trained to use their equipment, Barber-Greene was able to dominate the equipment market in the road-building and paving sectors in the postwar economy through the 1950s.

"It was used in the South Pacific and I think it was a Barber-Greene," recalled Peter Kemmetern, a retired Wisconsin highway official, speaking of an old asphalt plant. "After the war, one of the county governments in Wisconsin purchased the plant

Barber-Greeners
Lieutenant Don Wredling, Staff Sergeant John Halmagyi, and Lieutenant Frank Merrill, on the platform of an 848 Mixer set up for educational purposes at the Army Engineer School, Fort Belvoir, Virginia, in 1943.

from the U.S. government as a surplus item. It was just a beat-up old asphalt plant but the county guys loved showing it off. As it turns out, the plant was credited with a kill. The Seabees who were running the plant on an island in the Pacific came under attack by a Japanese warplane. The Seabees on duty making batches of hot mix shot the plane down and the credit for the kill went to the plant.

"The same way a bomber crew might have painted the national flag of the enemy planes they shot down under the pilot's cockpit window, the Seabees painted a flag of the Rising Sun on their asphalt plant. Years later, one of the state's contractors purchased the plant from the county and made sure to always paint around that Japanese flag. I wouldn't be surprised if the plant and the flag are still around."

Allies and enemy alike were impressed by the productivity and portability of the Barber-Greene equipment. As it turned out, their paving machines outperformed other nations' equipment by twofold, and were far more transportable.

Barber-Greene helped American forces pave asphalt runways, roadways, loading docks, and command-center facilities quickly and wherever they were needed. That alone was a crucial blow to the enemy and a significant contribution to the Allies' fight to save the world from tyranny. With this in the back of their minds, buying Barber-Greene equipment was almost a patriotic act for postwar asphalt contractors.

Opposite:

Aerial photo taken from the famous German airship *Graf Zeppelin,* shows the newly opened segment of the autobahn between Cologne and Dusseldorf. Inset, the German air force uses an autobahn as a runway. Junkers and Messerschmitt aircraft are hidden in the woods alongside the autobahn.

HITLER SHOWS OFF HIS HIGHWAYS

During the 1930s, Hitler's Third Reich invited highway engineers from around the world for a tour of the autobahns, the world's first superhighway network, that his minions were busy constructing. The head of the U.S. Bureau of Public Roads, Thomas H. MacDonald, along with other dignitaries and road builders, accepted the Fuhrer's invitation. Touring the road on the ground by automobile and in the air by dirigible left guests in awe—especially when German military vehicles sped by.

So impressed were they by the autobahn's style and efficiency, the members of the newly formed Pennsylvania Turnpike Commission decided it was time to build their own version of it.

Constructed quickly, between 1938 and 1940, the Pennsylvania Turnpike became known as the Granddaddy of the Turnpikes and boasted of its derivation from Germany's autobahn—including its concrete pavement and the lack of

speed limits. The turnpike's 160 miles of unencumbered superhighway were completed just as Hitler was turning his autobahns into a tool of war. Before the American public was allowed onto its first superhighway, U.S. Army truck convoys, foreshadowing world events, tested their vehicles' high-speed capabilities while traveling on a highway conceived on the autobahn model.

It would be seven years before another superhighway was built anywhere in the United States.

WORLD'S FIRST ASPHALT SUPERHIGHWAY

Maine shook off the dreariness of the war years by becoming the first state to build a superhighway in the postwar era. Opening to traffic on a cold December 13, 1947, just two years after construction began, the Maine Turnpike was just the second such roadway in the nation. Running between Kittery and Portland, the 45-mile structure differed from its predecessors in many ways but most remarkably in its material. The Maine Turnpike's surface pavement was black,

Opening day at the Maine Turnpike brought out a crowd of well-wishers, excited at the opportunity to drive this modern marvel.

not white—asphalt, not concrete. From its foundation to its travel surface, the Maine Turnpike was constructed entirely of asphalt.

The road was a major breakthrough for those in the world of asphalt who knew an asphalt highway could withstand the heavy loads inflicted on the roadway by trucking goods across Canadian–U.S. borders and endure Maine's arcticlike winters and hot summer days. Those who thought concrete was the only material for an autobahn-like road were suspicious. As with the autobahn, although minus the dirigible tours, highway engineers from around the world were invited to see Maine's asphalt superhighway for themselves . . . and they left impressed.

In a new era of streamlining goods and services in the public sector, the Maine Turnpike became a model of efficiency. The road was built for maximum use, serving the largest number of people for the least cost. As Roland Lavallee of consulting engineering firm HNTB recalls the details of the original paving, "The lanes were asphalt . . . but the shoulders were road gravel with a bituminous surface like a macadam." The state's choice to go with an asphalt highway was central to its immediate efficiency and future economy.

Echelon paving— using two pavers at once—helped to speed the construction process along Maine's turnpike. The wide medians were an innovative safety feature at the time.

The superhighway relieved congestion along the overwhelmed U.S. Route 1, which ran parallel. Route 1 had been a major source of tourism revenue since the 1920s, and the new road provided a far more efficient way for travelers and trucks to move through Vacationland, as the state bills itself. Lives were saved by the large median dividing its travel lanes and by its wide shoulders lanes that were nearly, but not quite, up to the standards of the future Interstate System.

Building on the successes of the original 45 miles of asphalt turnpike, the Maine Turnpike Authority began constructing a 66-mile extension in the spring

In 1956, 3,808,382 vehicles traveled Maine's asphalt. . . . and by 1986, 28,567,492 vehicle trips passed over the original asphalt base.

of 1954. More than 2000 construction workers began mixing 11,500,000 gallons of asphalt into 350,000 tons of hot mix, extending the legacy of the first super-highway in the world that was paved with asphalt.

In 1956, 3,808,382 vehicles traveled Maine's asphalt. Ten years later that number doubled, and it doubled again in the following 10 years and again in the subsequent 10 years, and by 1986, 28,567,492 vehicle trips passed over the original asphalt base. Today, the number of trips is more than 16 times the 1956 numbers.

THE MIRACLE TURNPIKE

The Minimalist Turnpike is more like it. The New Jersey Turnpike, called the Miracle Turnpike by some of the engineers who built it, was a 1950s model of project streamlining. Unique in its financing method of borrowing $230 million on an as-needed basis, this superwide, six-lane superhighway boasted a super-fast 75-mile-an-hour design speed and black-on-black construction—an industry term for a road with an asphalt travel surface and asphalt foundation. Despite the innovations, the project was completed in a record-breaking 24 months. If the Korean War had not caused a shortage of men and materials, the Turnpike Authority could have built its 240 bridges and 118 miles of road even faster.

"The turnpike has the distinction of having been financed, designed, and built faster than any comparable project ever has been," said *The New York Times* in its Sunday edition of January 20, 1952. The *Times* reported that the completion of the last nine-mile section brought an escape route to the city's "western doorstep." For the first time, citizens of the most populous city in America could drive out of their metropolis and out of their state by hopping onto a superhighway.

"The tortuous trip on the parallel road, U.S. No. 1 through Jersey City, Newark, and Elizabeth, is only an unpleasant memory for persons willing to pay an average of 1 and a half cents a mile for express highway travel over the 118 miles of turnpike," *Road International* magazine proclaimed in the spring of 1952, adding that the project was one of the largest asphalt highway projects ever and that it cut driving time between the Delaware and Hudson rivers from six hours to two. New Jersey was living up to its nickname, the Corridor State.

"Geographically New Jersey is one of the smallest states, but in terms of population and of importance to the economy of the country it is one of the greatest states. As a result of its location, it has a concentration of traffic the like of which does not exist anywhere else in the United States, or in fact, anywhere in the world," explained Paul L. Troast, the chairman of the newly minted New Jersey Turnpike Authority. Others were calling it "America's finest highway," claiming that it was "even surpassing the great Pennsylvania Turnpike."

Hyperbole or not, the swiftness of its construction and the durability of its pavements were undeniable. Speed of execution and quality of work were the Turnpike Authority's chief concerns during construction and its proudest accomplishments at project completion. All this while meeting and exceeding the future Interstate System standards—the turnpike was designed and built with 12-foot travel lanes, 10-foot-wide shoulders as "breakdown lanes" on the driver's right, and five-foot shoulders on the driver's left.

An early New Jersey Turnpike postcard showing the straightaway through Middlesex County. The card proclaims, "sight distances are long and curves are easy. There are no red lights and no left turns."

Everything on the job was rushed in order to meet the demands of investment bankers requiring assurances that the return on investment would be made as quickly as possible.

BLACK OR WHITE?

Civil Engineering magazine reported, "The first meeting of the engineering firms engaged by the New Jersey Turnpike Authority was called by Charles M. Noble, Chief Engineer of the Authority, in October 1949 and was attended by all the section engineers, the consulting engineers and the Turnpike engineering staff. The question uppermost in everyone's mind at the time was whether the Turnpike would be black or white."

Experts understood that either surface could endure the heaviest traffic conditions in the country. The nub of the issue was which one could do it for the least cost up front and a minimum maintenance budget over the life of the road. Minimizing the short-term and long-term costs became the deciding factor. To address the question, a pavement committee was formed. Members included representatives of some of the best engineering firms in the nation and key members of the Turnpike Authority.

The pavement committee laid out a possible solution when it considered specifying concrete for the southern sections and asphalt for the northern ones, and accepting bids for both pavement types in the middle of the state.

The single 118-mile megaproject was broken down into seven construction contracts. The pavement committee laid out a possible solution when it considered specifying concrete for the southern sections and asphalt for the northern ones, and accepting bids for both pavement types in the middle of the state. They specified asphalt for the northern portions of the road because that area traverses the unstable and unpredictable marshes at the Meadowlands. Concrete, it was feared, would break apart if the highway should settle in the swampy area.

Specifications for both a "rigid" concrete surface and a flexible surface were created. Keeping the black-or-white contest fair, design specifications for the rigid pavement were drawn up in part by the Portland Cement Association, and flexible pavement criteria were reviewed by the Asphalt Institute and the Army Corps of Engineers.

In a significant departure from its plans, the Turnpike Authority eventually decided to go for an all-out bidding war. Their logic: either material would be hard-wearing. The lowest bidder on each of the seven sections would ultimately answer the question of black or white.

Stones form the solid foundation of the New Jersey Turnpike. The structure of this superhighway is so strong, it never had a structural failure in its first 50 years, and was recognized with the first-ever Perpetual Pavement Award in 2001.

LET THE BIDDING BEGIN

As it turned out, concrete didn't stand a chance. Asphalt prevailed because it required less labor and lower-cost materials. Black or white, the superhighway was going to be built with identical 17.5-inch subgrades and nearly identical six-inch subbases that together made up two thirds of the total mass of the three-foot-thick road. The asphalt specifications called for a travel surface with 4.5 inches of hot-mix pavement over an 8-inch macadam base. The concrete surface would be a 10-inch-thick layer of expensive, labor-intensive, steel-reinforced concrete.

The specification for the New Jersey Turnpike's asphalt travel surface called for three layers of hot mix, each 1.5 inches thick. The final design specification changed the 7.5-inch waterbound macadam base to the same thickness of "Asphalt Penetration Macadam Base," to be laid in two layers; the first was to be 4.5 inches thick and the second one 3 inches thick. Even when combining the 4.5- inch depth of the asphalt travel surface with its 7.5-inch asphalt penetration macadam base for a total thickness of 12 inches, asphalt was still economically superior—$5,457,134 superior.

The total bids for concrete pavement came to $45,729,568, or approximately 20 percent of the $230 million total job. The winning bids for an asphalt highway totaled $40,272,434. Regardless of their understanding of the value of $5,457,134,

Rolling stone prior to applying asphalt on northbound lanes south of Route 40.

the investment bankers who had the last say in all fiscal matters had to be convinced that asphalt would hold up. After all, it was their investment foremost. They were not disappointed.

As New Jersey was making its final decisions on black or white, Maine had some encouraging news on black highways for their fellow turnpike authority. After tallying up its maintenance cost for the first four years of operation, the bill came to just $12,000—only $64 per year per mile for the four-lane, 47-mile-long route. This in a state where brutal weather puts any road surface to the test!

Looking ahead, the engineers projected that later in the New Jersey Turnpike's life, replacing sections of an asphalt roadway would be far less involved than replacing slabs of concrete. Concrete-slab replacements would require more downtime, whereas an asphalt roadway could be up and running sooner with fewer detours and far less negative impact on toll revenue.

Finally, asphalt's capabilities showcased during World War II were a fact. Herbert Spencer, an asphalt engineer, explained, "The wartime history of asphalt pavements provides an outstanding service record. Access roads and airfields built in the United States and all over the world have stood up under the most grueling traffic. At the General Motors proving grounds, asphaltic concrete was the only pavement that successfully withstood the shearing, tearing action of 50-ton tanks."

Spencer reminisced, "One of the most eminent road engineers in the United States at the turn of the present century was James Owens of Essex County, considered one of the pioneers in use of macadam roads. The stone bases and Telford foundations he built fifty years ago still serve as adequate foundations, indicating the value of granular materials for the support of the loads imposed."

Finally, it was official: asphalt prevailed in all seven contracts. The chairman of the New Jersey Turnpike Authority, Paul L. Troast, commented, "After careful consideration of all factors involving pavement, such as cost, durability and safety features, the Commissioners of the New Jersey Turnpike Authority have decided to use asphaltic concrete for the pavement of its 118 mile super-highway." He added, "The Turnpike will be superior to any road in existence."

A BLACK FUTURE

In January of 1950, work on the New Jersey Turnpike began. No one knew that the giant project was a harbinger of the next 50 years of Interstate System construction. Requiring the coordination of 700 engineers and 10,000 construction workers, it was New Jersey's largest public works project ever.

The start of construction on the gargantuan road project drained the available resources of the region's dredgers, cranes, excavators, earthmovers, trucks, work trailers, asphalt plants, pavers, and power rollers. Insurance adjusters put the replacement cost of the hardware at $45 million, an incomprehensible sum to some. Contractors from Philadelphia, New York, and New Jersey made up the army building the road.

Convoys of trucks rolling up and down the alignment reminded more than a few veterans of the "war period." Mammoth machine shops ran night and day to keep gear moving while crews roaming the field serviced the fleet during the 24-month job.

Asphalt paving began in the spring of 1951. It was completed that fall—fast and black.

On a single job site, as many as 17 batch plants were supplying 17 Barber-

Asphalt paving began in the spring of 1951. It was completed that fall—fast and black.

A hungry beast, the New Jersey Turnpike required enormous asphalt plants to produce the material required for the huge volume of paving.

Greene and two Adnun pavers, and it was not unusual to see as much as 175 tons of asphalt being laid per hour. Immediately following laydown, 10-ton power rollers compacted the hot mixes. With such intense activity and enormous tonnages of asphalt being laid down, field inspectors reached the limit of their ability to enforce mix designs.

Overall requirements were 14 million gallons of asphalt binder for the penetration macadam base, 11.3 million gallons of asphalt for the hot-mix, 870,000

tons of aggregate and 15 large asphalt plants to do the hot-mixing. On a daily basis that meant an average laydown of 15,000 tons of aggregate and the mixing of 160,000 gallons of asphalt. Tanker trucks hauling the liquid asphalt to the job came from oil refineries in Pennsylvania and New Jersey.

Not all contractors who won the bids to pave understood the properties and behaviors of asphalt, and not one of them had ever worked on an asphalt-paving job this large. In an unusual show of unity, the bid-winning contractors met for

a briefing in Trenton, New Jersey, before the paving began. Contractors with more experience assisted those with less understanding of the handling and application of large volumes of asphalt.

All seven sections of the turnpike's paving project had similar but different gradations and therefore different but similar requirements in their hot-mix designs. The road surfaces were to have a range between 5.25 pounds and 5.8 pounds of asphalt mixed into each 100 pounds of stone used to create the asphalt travel surface

The strength of the asphalt was to be determined by the Marshall test. The Turnpike's design called for it to meet a 36,000-pound-axle-load requirement—twice the amount of weight allowed on most turnpikes at the time. The stones laid down on the three-foot-thick road would be compacted up to 95 percent of possible density. The Army Corps of Engineers offered input and the Asphalt Institute was heard from, injecting its views and opinions.

Originally, the New Jersey Turnpike Authority required that the asphalt travel surface be laid down in two layers, or "lifts." The authority's original specifications required side forms during paving. Asphalt paving contractors viewed this as a holdover requirement from prewar paving methods. Seeing a better way, they launched a protest with the paving committee.

Knowing the speed at which they could perform the required work with a Barber-Greene floating screed, the paving firms on the job objected to the restraining requirements. The contractors prevailed in their arguments with the Turnpike Authority with two conditions: First, the four-and-a-half-inch asphalt travel surface would be laid in three thinner layers instead of just two lifts. Second, the final driving surface would be required to be as smooth as or smoother than if side forms had been used. An agreement between the paving committee and paving contractors was reached, and the side form requirement was dropped.

Savin Construction, after winning its bid for asphalt paving, requested another change.

When manufacturing the base of the asphalt travel surface, Savin suggested that changing the specifications from eight inches of waterbound macadam to 7.5 inches of asphalt penetration macadam would benefit the turnpike in the long run. An asphalted base, Savin argued, would make for a stronger and more durable foundation for the asphalt travel surface. As other contractors won their bids, they too asked for the same change to their contracts.

Again, the Turnpike Authority agreed; and contractors, when laying out the

The New Jersey Turnpike's design called for the pavement to be able to bear 36,000 pounds per axle load—twice the amount allowed on most turnpikes at the time.

two layers of asphalt penetration macadam, poured as much as 3.2 gallons of asphalt over each square yard of gravel. Asphalt sprayer trucks like the ones used to lay down Tarvia in an earlier era were employed for the job. The benefits were seen sooner than expected.

When completed, two layers, one 4.5 inches and the other 3.5 inches thick, both heavily coated with asphalt, made for an excellent temporary haul road during construction. By providing a way for construction equipment to move about the sandy and swampy areas, the enhancement in mobility sped up construction.

Complications with the turnpike's drainage systems arose during construction, threatening an adverse impact on the completion schedule. The asphalt penetration macadam layer became a saving grace because it sped the movements of teams working on the problematic drains. As a result, the Turnpike Authority was able to resolve the threat to the schedule.

The turnpike reflected the legendary thickness of the Roman road—it too was three feet thick—with two feet of crushed rock (waterbound macadam) supporting an additional foot of asphalt in the form of 7.5 inches of asphalt pene-

Quality inspections were part of every phase of the job, ensuring that all work on the turnpike was performed according to specifications. Sometimes it seemed that an army of white-shirted inspectors was hovering over the paving crews.

tration macadam and 4.5 inches of asphalt travel surface.

At the peak of paving, there were 15 hot-mix plants along the 118-mile alignment, mixing the asphalt needed for the 6,300,000-square-yard job. Showing the diversity of the plant manufacturers at the time, the plants working on the turnpike were manufactured by Warren Brothers, McCarter, Madsen, Standard Steel, Simplicity, and Cummer.

The Tioga Paving Company ran four plants in one locale, creating a monster—a single super-megaplant with an average daily output of 4000 tons per day. Tioga's behemoth plant often put out 5000 tons a day and even exceeded a record-breaking 6000 tons on one workday.

CONTROLLING QUALITY

Contractors of the day marveled at the state-of-the-art push-button controls for mixing the average 120 tons of hot mix per hour. With automation came increased production and a better product.

CHECK YOUR GAS
CITIES △ SERVICE
LAST GAS FOR
26 MILES

The Woodbridge interchange was an engineering marvel of its time. The curves were gentle and access was controlled, so there were no worries about what might happen at the next intersection.

Left, travelers unfamiliar with the route found driver-friendly signs advising them of opportunities to stop for necessities such as gas.

Top right, ten-foot-wide "breakdown lanes" on the driver's right enhanced safety and allowed traffic to keep moving if a car had trouble.

Washington Bridge   U.S. 46   N.J. 3   Secaucus   Lincoln Tunnel

WEATHER WARNING

NO
HOUSETRAILERS

EXIT ☐ TO ☐

INTERCHANGE 11
WOODBRIDGE–THE AMBOYS   WIND

Even after the initial construction was complete, the New Jersey Turnpike continued to set records. The 24-lane toll plaza at the Lincoln Tunnel Complex was the world's largest at the time. Right, signs on toll booths were designed to warn motorists of hazardous conditions ahead. Left, the turnpike's service areas set a high standard. Attendants pumped gas, cleaned windshields, and checked oil routinely. A motorist could even have the oil changed while dining in one of the adjacent restaurants.

Each of the 15 asphalt plants had its own field laboratory and teams of quality inspectors. Looking more like a M.A.S.H. unit's tent from the Korean War than a quality control center alongside one of the most sophisticated road-building projects to date, the inspection unit had personnel checking "the method of feed, the capacity of dryer, screens and aggregate bins, the automatic controls of the weigh box, and the pug mill." Once the inspectors found the plant to be in good working order, the contractor was allowed to begin mixing.

While inspectors checked the plants for performance, other field laboratories and personnel inspected the hot mixes coming out of the plant. Other trailers were converted into field laboratories to solve problems associated with consistency of mixes. With so much asphalt hitting the road and so little experience with laying the vast quantities needed, the more eyes on the material the better.

After the pavement was laid, an independent inspection firm drilled into the new superhighway and removed a core sample. The test section was then inspected at a laboratory and either rejected or approved.

When it came to smoothness, a one-eighth-inch variance for every 10 feet of travel surface was the maximum allowed. In other words, nearly none. One of the least favorite sights of the paving contractor was the man with the roughometer, an I-beam on wheels with bell and marker measuring the roughness of the road.

Speed of fundraising, designing, building, and paving became the legendary road's trademark. The New Jersey Turnpike and its authority had made history with its large and successful application of asphalt. More significant, the turnpike's 118 miles of innovative ways would set the pace for the construction of the nation's future 40,000-plus miles of Interstate System.

The New Jersey Turnpike's 118 miles of innovative ways would set the pace for the construction of the nation's future 40,000-plus miles of Interstate System.

"ROADS NOWHERE"

Many travelers remember marveling at the new turnpike as they drove along it, perhaps with the top down, the radio on, and enjoying the fast ride when suddenly for no obvious reason the new road came to an abrupt end. They were unceremoniously dumped onto an old U.S. route, fighting traffic lights and traffic jams.

When the Pennsylvania Turnpike was first built, it dead-ended into Ohio, where the Ohio Turnpike had not yet been built. The Maine Turnpike's 47 miles of highway dead-ended into New Hampshire, whose short but critical turnpike had not yet been constructed. This meant that vacationers pouring out of Maine on the new superhighway suddenly were crawling along old U.S. Route 1 in New

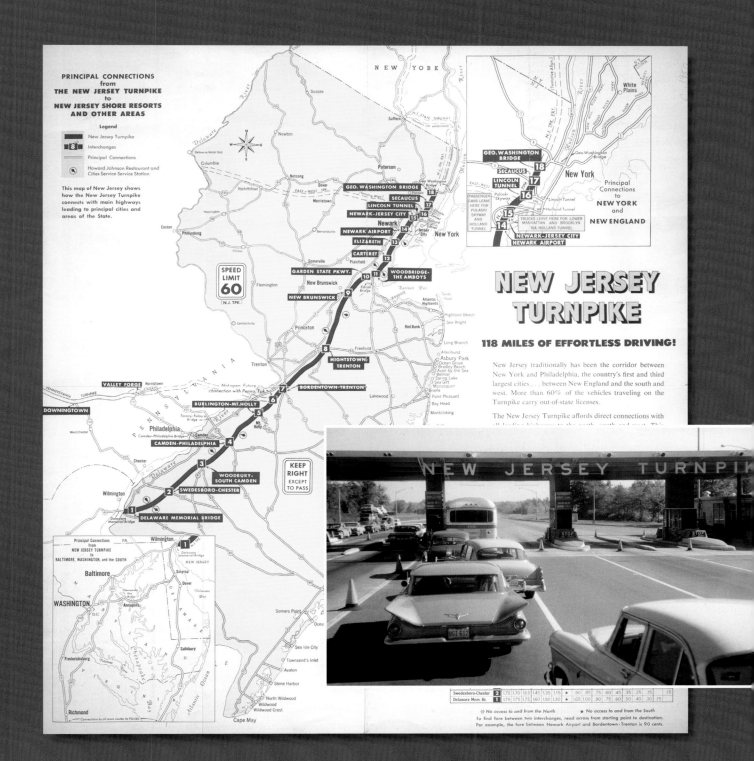

PRINCIPAL CONNECTIONS
from
THE NEW JERSEY TURNPIKE
to
NEW JERSEY SHORE RESORTS
AND OTHER AREAS

Legend

New Jersey Turnpike

Interchanges

Principal Connections

Howard Johnson Restaurant and
Cities Service Service Station

This map of New Jersey shows
how the New Jersey Turnpike
connects with main highways
leading to principal cities and
areas of the State.

SPEED LIMIT 60 (N.J. TPK.)

KEEP RIGHT EXCEPT TO PASS

Principal Connections
to
NEW YORK
and
NEW ENGLAND

NEW JERSEY TURNPIKE

118 MILES OF EFFORTLESS DRIVING!

New Jersey traditionally has been the corridor between New York and Philadelphia, the country's first and third largest cities . . . between New England and the south and west. More than 60% of the vehicles traveling on the Turnpike carry out-of-state licenses.

The New Jersey Turnpike affords direct connections with

Principal Connections
from
NEW JERSEY TURNPIKE
to
BALTIMORE, WASHINGTON, and the South

Swedesboro-Chester	2	1.75	1.70	1.65	1.45	1.35	1.15	★	.90	.85	.75	.60	.45	.35	.25	.15		.15
Delaware Mem. Br.	1	1.75	1.75	1.60	1.50	1.50	★	1.05	1.00	.90	.75	.60	.50	.40	.30	.15

※ No access to and from the North   ★ No access to and from the South
To find fare between two interchanges, read across from starting point to destination.
For example, the fare between Newark Airport and Bordentown-Trenton is 90 cents.

Hampshire. And in 1952, the New Jersey Turnpike dead-ended into the state of Delaware. These bottlenecks were not a conspiracy against drivers, they were just signs of the times.

In the late 1940s and early 1950s, the federal government was preoccupied with demobilizing its military forces while restoring the world's balance of power and then launching a war in Korea. Road-building policies at the federal level drifted as the states, in a series of fits and starts, began a haphazard, self-motivated, and uncoordinated attempt to build their own superhighways—with little or no regard for what neighboring states were doing. Without a central governmental body leading, coordinating, and executing a master plan, the disarray of a national road network was the rule just as U.S. citizens were demanding something better.

In the 10 years after World War II, there were 45 different turnpikes, toll roads, and pay-as-you-go highways in various stages of planning, construction, or use. As of January 1955, the federal government tracked the shortest toll road, the New Hampshire Turnpike, at 15 miles, and the longest, the New York Thruway, at 396 miles.

Independently conceived and operated turnpikes meant little or no cohesion along America's growing but still infant superhighways. In the years after World War II, only the prosperous states that could afford to raise funds and promise repayment to bondholders built roads, while the less affluent or less influential states suffered with dangerous, overburdened, and broken-down thoroughfares. Looked at from a national level, highway planning and construction drifted into a disorganized, inequitable period of building and paving.

Thus, engineering marvels or not, the early turnpikes were, more often than not, roads to nowhere. Nearly all of the new turnpikes lay in the Northeast and only a few west of the Mississippi River. Not until nearly 10 years after World War II did any two of them link up to make a single interstate superhighway.

In 1955, *The New York Times* reported that the Ohio Turnpike, in "its joining with the Pennsylvania Turnpike last December 1, marked the first physical connection of two major toll routes and foreshadowed the creation of a vast superhighway system in the East. Barring unexpected hitches, it should be possible to drive nonstop between New York and Chicago over the system by this time next year, paving the way . . . to go one-third of the way across the nation without stopping for a traffic light or an intersection at grade. . . . Only two gaps now remain in the 828 miles of turnpike pavement connecting the two cities."

The "vast superhighway system in the East" left the rest of the nation want-

Opposite:
The Ohio Turnpike, about seven miles west of Youngstown. At the top left is Meander Creek Reservoir, one of many scenic spots along the turnpike. When Ohio linked its turnpike to Pennsylvania's on December 1, 1954, it was the first time any two state's turnpikes had joined together to form a single interstate superhighway.

ing. Even the heralded "super road" needed to fill two gaps in the exclusive five-state interstate turnpike system: Indiana's 157 miles of toll road needed to be completed and a massive bridge was required to link the New Jersey and Pennsylvania turnpikes.

At first the West Virginia Turnpike was called "the road that couldn't be built." Once it was built, it was widely known as the "Road Nowhere" with newspaper reports saying, "On a map this toll highway gives the appearance of a road that leads to nowhere." The West Virginia Turnpike authorities in 1954 had expansion plans that were not entirely their own to decide. They stated hopefully that "The first step on extension will be southward from Princeton to the Virginia border, but this is awaiting completion of surveys by the Old Dominion [Virginia] Turnpike Authority to determine the most feasible route through Virginia."

Farther west, in Kansas City, on October 21, 1956, cowboy movie star Gene Autry jumped his horse Champion through a giant map of the Kansas Turnpike, heralding the opening of their portion of the multistate highway and delivering a superhighway that met the design standards of the recently announced Interstate System. Neither the presence of the cowboy star nor the fact that the sitting President of the United States at the time, Dwight D. Eisenhower, was from Kansas, precluded that state from the dishonor of its own embarrassing dead end.

Building a multistate turnpike that would link the Gulf of Mexico with the Great Lakes was the brainchild of Governor Johnson Murray of neighboring Oklahoma. The problem was that by the time Kansas planned and built its sections of Murray's megaroad, Oklahoma still hadn't raised the money to get started, leaving the Kansas Turnpike to dead-end into the fields of an Oklahoma farmer named Amos Switzer.

High-speed neophytes, who had never seen a superhighway let alone driven on one, would barrel south down the newfangled Kansas Turnpike and overshoot its dead end.

High-speed neophytes, who had never seen a superhighway let alone driven on one, would barrel south down the newfangled Kansas Turnpike and overshoot its dead end. Vehicles crash-landed into Amos's fields at the alarming rate of one a day, making it difficult for him to do much aside from rescuing marooned motorists and attending to the injured and bewildered.

The kindhearted Amos set to plowing his field for softer landings and patching the holes in his fences created by the wayward vehicles. Finally, when Wyoming's governor, Millard Simpson, and his first lady took the infamous leap into the fields, bringing unwelcome fame to both farmer and chief executive, Amos had had enough. Besides, he was having to sell some of his livestock due to the holes in his fence.

Blessing or curse?
Oklahoma farmer Amos
Switzer had a super-
highway sprouting out
of his wheat field. While
he may have had an
inside track on getting
his crop to market,
there was a downside—
cars kept winding up in
his field.

The now angry agrarian requested that Oklahoma erect a large wooden bar-
ricade at the end of its turnpike. Within 24 hours of its completion, however,
three cars careened through it, reducing it to splinters. Five hundred crash landings and a year and a half later, the Kansas Turnpike—and Amos's fields—were
converted into I-35, making the badly needed connection.

Amos's dilemma created a textbook case for why the federal government, not
48 independent states, needed to orchestrate the building of Ike's Grand Plan.

CHAPTER THIRTEEN

Ike's Grand Paving Plan

"Americans are living in the midst of a miracle. A giant nationwide engineering project—the Interstate Highway System—is altering and circumventing geography on an unprecedented scale. Rearranging the geography of America, the revolutionary Interstate Highway System creates a 41,000-mile network of super-roads from border to border. Yet an individual can no more grasp the scope and magnitude of the entire project than an ant can comprehend New York City . . . [It is] perhaps the greatest revolution in ground transportation since the wheel."

232

IN JUNE OF 1945, America's most triumphant general of its largest war effort ever returned home. The victorious Supreme Commander of Allied Forces and an armada of warplanes first touched down in Washington, D.C., before heading to New York City for a ticker-tape parade. On Capitol Hill, Senators, Congressmen, and anybody who was anybody packed into the nation's Capitol building to hear the man speak. After a heartfelt speech to a war-weary audience, the five-star general received the longest standing ovation in Congressional history. In a town where people disagree for the sake of it, everyone seemed to agree, "I like Ike."

More than seven years later, Ike returned to Washington. On January 20, 1953, Dwight David Eisenhower was sworn in as the 34th President of the United States. Transforming his hatred of war into a relentless pursuit of happiness for all Americans, the President made the construction of the Interstate System his favorite item on the domestic agenda. Eisenhower knew that superhighways could tie the nation together—socially, economically, and militarily.

Transforming his hatred of war into a relentless pursuit of happiness for all Americans, the President made the construction of the Interstate System his favorite item on the domestic agenda.

Prior to Ike's arrival in Washington, the plan was to build a 40,000-mile interstate system by improving and upgrading the most heavily used routes. The likes of U.S. Route 66 and U.S. Route 1 would be improved piecemeal: reducing the grade of a highway here, straightening a sharp curve there, and creating a divided highway now and again, as time and money permitted. For Ike, this approach was far too slow.

As conservative as his politics were, Ike believed that a centralized body, the federal government, not 48 individual state governments, should finance the new interstate highway system. He couldn't bear to see the states, and every American, languish in a political quagmire of indecision over budgets and rights of way. Ike wanted the federal government to work with the states in designing the routes and then reimburse the states for their work. Historically, Rome, France, and Germany had built the finest road systems in the world from the top down and so would the United States—with his leadership.

Ike's Grand Plan was to upgrade the nation's entire road network over a 10-year period. He envisioned a new highway system that resembled the autobahns of Germany. Instead of simply improving the old U.S. routes, his plan would result in building a brand-new system that would replace the old, often running parallel to and within sight of the highways.

IKE BUILDS HIS CASE

In 1949, the nearly 40,000 miles of highway slated to become improved interstate highways had been audited by the Public Roads Administration (PRA) and found to be in poor shape. The average age of a road surface was 12 years. Nothing was consistent or uniform about the roads. Drivers could never guess when a wide roadway would suddenly narrow. Anticipating the rise and fall of the road or the sharpness of a blind curve was impossible. A bridge might hold your truck's weight, or it might not.

The PRA estimated that it would take 20 years to upgrade the roads to standards acceptable for the demands that were currently being put upon them.

Twenty years was too long for Eisenhower to wait. Over three consecutive years of his presidency, Ike used his State of the Union addresses to win support from the public and rally the members of Congress. On January 7, 1954, he said we were about to engage in "the building of a stronger America. . . . [T]o protect the vital interests of every citizen in a safe and adequate highway system, the Federal Government is continuing its central role . . . so that maximum progress can be made to overcome present inadequacies of the Interstate Highway System."

Ike strategically chose the July 12, 1954, Governors' Conference, an annual meeting for the country's governors, to roll out his Grand Plan. Sadly, Ike's sister-in-law passed away, and he was prevented from attending, but he left prepared remarks for his Vice President, Richard Nixon, who faced an unexpectedly hostile crowd. When it came to the federal government and road building, many state governors were telling the feds to get out of the road-building business altogether. The governors were stunned by Ike's message.

Ticking off the costs of poor roads, the speech pointed out that the death rate was "comparable to the casualties of a bloody war," as 40,000 people a year were killed on the highways. Civil suits were clogging the courts and economic losses were in the billions of dollars because of highway inefficiencies, detours, and traffic jams. Most compelling of all were the highways' "appalling inadequacies to meet the demands of catastrophe or defense, should an atomic war come." Although the new highway system would be the largest project in the history of the world, Ike believed the highways would pay for themselves.

The Interstate System would help to eliminate dangerous grade crossings such as these.

The room full of governors was buzzing. Nixon concluded, "I would like to read to you the last sentence from the President's notes, exactly as it appears in them, because it is an exhortation to the members of this Conference. Quote: 'I hope that you will study the matter, and recommend to me the cooperative action you think the Federal Government and the 48 states should take to meet these requirements, so that I can submit positive proposals to the next session of the Congress.'"

In other words, let's build it together! It was too good to be true, or so many of the governors thought. But the governors saw an opportunity. The speech broke the ice for Ike's Grand

By the early '50s, road congestion was a factor of daily life for commuters and the Interstate System was seen as a solution.

Plan for a new Interstate System. With the question out of the way of who was going to be in charge, and with what to build coming into focus, the next big question would be: Who is going to pay for it?

Ike reiterated his concerns in his State of the Union message of 1955: "A modern, efficient highway system is essential to meet the needs of our growing population, our expanding economy, and our national security." It was a year that Ike was consumed with the potential for a nuclear strike from our enemies.

If nuclear bombs started dropping, it was estimated that 70 million people would need to be evacuated from urban areas, and the military was certain the best way was over superhighways. The army wanted an interstate system so it could move troops into the strike zone. It requested beltways around cities, enabling rescue personnel and equipment to bypass congested areas while rushing to a disaster zone. As a result, in 1955, 2300 miles of such bypass roads were designated for the future Interstate System.

A LONG SHOT

The Republicans lost their control of the House and Senate in 1954, making a long shot out of pushing legislation through Congress. Both parties supported the highway legislation, but coming up with an acceptable financing plan proved difficult. Political skirmishing over highways occupied most of 1955. Complicating matters even further, Ike suffered a serious heart attack in the fall of 1955. Regardless of the public's desire and the nation's needs, 1956 didn't look good for Ike's Grand Plan.

In a letter dated February 22, 1955, the President delivered to the American public a message detailing his plan. He emphasized that "Each year more than 36,000 people are killed and more than a million are injured on the highways. . . . In case of atomic attack on our key cities, the road net must permit quick evacuation of target areas, mobilization of defense forces . . . " In addition, he said, the nation's gross national product was threatened by traffic jams.

In January of 1956, the Eisenhower administration delivered the Annual Economic Report to Congress. It pulled out all the stops in order to get the Grand Plan under way, saying, "The country urgently needs a modernized interstate highway system to relieve existing congestion, to provide for the expected growth of motor vehicle traffic, to strengthen the nation's defenses, to reduce the toll of human life exacted each year in highway accidents, and to promote economic development."

Members of the President's advisory committee on national highways are shown as they presented President Eisenhower with its recommendations. Left to right, standing, General Lucius D. Clay, chairman; Frank C. Turner, executive secretary of the group; Stephen D. Bechtel, of San Francisco, California; S. Sloan Colt, of New York City; William A. Roberts, of Milwaukee, Wisconsin; and David Beck, president of the International Brotherhood of Teamsters.

THE BATTLE WON

Despite Ike's heart attack and the Republicans' loss of control of Congress, 1956 was the year everyone came together: Tire, automobile, truck, and petroleum manufacturers joined with construction and travel associations. The White House and Congress, Republicans and Democrats saw the benefits of the Grand Plan and their own stake in it. Totally engaged and taking nothing for granted, the entire bunch threw their support behind the Federal Aid Highway Act of 1956.

Congressman Hale Boggs of Louisiana devised a federal Highway Trust Fund that gave the bill real strength in funding. The federal government would collect money from user fees such as a gas tax, providing the money to pay the states back 90 percent of the cost of constructing the Interstate System. The House passed the bill 388 to 19, giving it an astounding 95 percent approval. The Senate, receiving the bill from the House, approved the legislation with a simple voice vote.

On June 29, 1956, while recovering from an intestinal infection and surgery at Walter Reed Hospital, Ike signed the bill into law. With no fanfare, media, or spectators, he changed the course of American history.

Ike's Grand Plan was under way, and the American public was happy about the news. Heaping credit on the President, they reelected him in 1956 with 57 percent of the popular vote—the highest ever bestowed on a presidential candidate in American history. His first term saw the start of the largest paving project in the world and his second term would see the completion of nearly a quarter of it.

Senator Dennis Chavez, New Mexico, chairman of the Senate Public Works Committee, celebrates after the passing of the highway bill. The legislation created the Highway Trust Fund, which funneled user fees into the highway program.

MR. ASPHALT

Mr. Asphalt, a.k.a. Sheldon G. Hayes, was 35 years of age when he purchased the first Barber-Greene floating screed for commercial use. When it was delivered in 1930 to Hayes's Route 66 town of Rolla, Missouri, the Barber-Greene Company couldn't have found a better promoter for their futuristic machine.

In an industry full of colorful characters, Hayes stood out. Having started Cadillac Asphalt Paving Company in 1946 in Detroit, Hayes became the man to know in Michigan

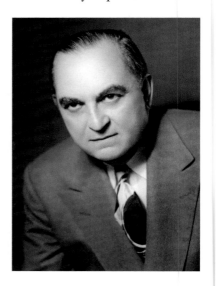

Sheldon G. Hayes, right, "Mr. Asphalt" at the time he built the Flatrock Speedway.

Sheldon G. Hayes mans the roller as flagman and Detroit Lions football player Leon Hart waves him across the finish line in 1952. The track opened the next year. Hayes regularly employed as many as twelve players from the Lions, who were making only $5000 to $6000 a year playing football. They liked the paving business because it helped them keep in shape in their off season.

Hayes, standing center, examines the special "steel asphalt" mix used on the Speedway. Scott Baker is at far right.

when talking asphalt. One of Hayes's pet projects was the Flatrock Speedway, a quarter-mile track with a standard stone base. The tight turns of the short track guaranteed excitement, and the stone base did an effective job of draining runoff from rain into the infield. So excited about the work was Hayes that he "was down at the speedway every day, driving around in his Cadillac and watching things," recalled his son, Sheldon W. Hayes.

Hayes had a passion for asphalt and a flair for organizing—to such an extent, he energized his fellow contractors to join with him in forming the Michigan Asphalt Paving Association in 1950. At that time, the Asphalt Institute was the only national organization representing the asphalt industry. The Institute traced its roots to the Asphalt Association, which had been formed in 1919 by a group that included both refiners and contractors. The organization became highly influential on technical and engineering issues, but by 1929, the contractor members thought there should be more emphasis on marketing and promotional efforts. Unable to resolve the two camps' differences, the Asphalt Association disbanded on December 31, 1929, and on January 1, 1930, a new organization, the Asphalt Institute, was formed with only refiners of asphalt cement as members.

Beginning in the 1930s, hot-mix asphalt producer/contractors in a number of states had begun forming statewide associations to represent their interests. Pennsylvania was the first, in 1932, followed by New Jersey in 1933, Massachusetts in 1935, Illinois and Kentucky in 1938, and Ohio in 1943. By 1955, there were 20 state asphalt pavement associations, but producers of hot mix still did not have their own national association. With the Interstate System on the horizon, Hayes and his fellow contractors knew it was time to get better organized at the national level. In 1954 and 1955, they began meeting to discuss forming an organization that would represent their interests exclusively. The staff leaders of the state associations agreed that the challenges of the time demanded that efforts be coordinated through a national association.

Much was at stake. Concrete producers, represented by the well-organized and well-funded Portland Cement Association and the American Concrete Pavement Association, were pouring millions of dollars into a campaign to pave the future Interstate System white. The largest paving job in the history of the world was about to get under way, and asphalt producers and pavers feared that they would be left behind by concrete producers whose campaigning was giving them a leg up in the debate on concrete versus asphalt.

On May 17, 1955, the national association's birth certificate, in the form of a legal charter for incorporation, was filed. It was decided that the organization should be called the National Bituminous Concrete Association (NBCA) because about half of all the existing state organizations used "bituminous" in their names. The initial logo was a magic carpet of hot mix. Ten years later, recognizing the lack of public understanding of the term "bituminous concrete," the association would change its name to National Asphalt Pavement Association (NAPA).

The minutes of the first meeting of the Board of Directors stated, "The boys feel that a National Association should work along with the Asphalt Institute and together they should do the industry some good."

One of the presentations at the first annual meeting of the newly created association was given by Captain Charles M. Noble of the New Jersey Turnpike Authority. His lecture was titled "Bituminous Concrete on the New Jersey Turnpike." In a crowded room of contractors at Chicago's Conrad Hilton Hotel in 1956, it was a hot topic.

Not all issues at NBCA were about the pending Interstate System. One of the first affairs requiring a rallying of the troops was to encourage an all-out legislative assault on the United States Air Force's decision to use only concrete on its runways. The battle over the runways, not the highways, would be the new national asphalt association's first legislative victory. (*See* chapter 11, "Paving the Way to Victory," page 198.)

With the battle of the runways behind them, asphalt producers set their sights on the land battles of the new Interstate System. To that end, NBCA launched a research program, announcing it at the fourth annual meeting, in 1959. At the same time, Charles R. Foster, who had been chief of the flexible pavements branch at the Army Corps of Engineers' U.S. Waterways Experiment Station in Vicksburg, Mississippi, was hired as the NBCA's technical coordinator. Foster had been part of the team successfully testifying before Congress that asphalt was well-suited to airfield pavements.

The Quality Improvement Program was the first of the fledgling association's initiatives aimed at expanding the knowledge base about asphalt, educating members of the industry, and enhancing the quality of the pavements that were designed and constructed in the United States.

In 1960, after a national search during which a number of university locations were considered, the Quality Improvement Program found a home at Texas A&M University in College Station, Texas. The program was funded separately from the NBCA's other activities, and the association's members donated generously to support it.

Named trail to superhighway: The old Mark Twain Highway in Missouri became U.S. Route 40 in the 1930s. Its next transformation started on August 13, 1956, when Missouri started to turn this section into Interstate 70.

THE INTERSTATE DECADE: 1956–1966

Americans marveled as over 1.6 million acres of land was set aside for the modern miracle of a national network of superhighways. It was enough land to cover Rhode Island twice over. Each mile of highway being built required an average of nearly 40 acres of land. To build all 41,000 miles required moving enough earth to cover the state of Connecticut knee-deep in dirt. The sand, gravel, and stone going into the Interstate System was enough to build a wall 9 feet thick and 50 feet high around the globe. Legend had it you could see the new highways from the moon.

The entire Interstate System was originally scheduled for completion in the fall of 1972. That proved to be too ambitious a deadline. Instead, the

work dragged on into the next millennium. However, between the years 1956 and 1966, more miles of the Interstate System were constructed than in any comparable period in its history. More than half of the system was built during the Interstate Decade. And what a show it was.

"In 1956, I don't think anybody had any idea what was ahead—inside or outside of the road building industry," reflects Mike Elliott, an industry leader. "But it didn't take long for them to realize just how enormous this job was going to be," he adds.

"America was humming with the sound of heavy equipment. Giant earthmovers stole the shows in towns across the country. Following them were crews building the road's subbase, its base and intermediate grades, and its finished courses—the paving. It was a tremendous show in every town it was being built in.

"Very few citizens had any notion or grasp of how this Interstate System was going to affect their lives," continued Elliott. "People were myopic, looking as far as their backyard or maybe the next county over, but no one thought much about what America would be like when the massive web of highways was completed—what it was going to mean to the average Joe in the rural areas or the millions packed into the cities."

With amusement and amazement Elliott remarked, "Starting out, land appropriation was the biggest task—procuring the right of way for over 42,000 miles of superhighway was an incomprehensible undertaking."

PRESIDENTIAL PAVEMENT

At the height of the Interstate Decade, President John F. Kennedy showed his support for his predecessor's Grand Plan by cutting the ribbon of a new section of Interstate 95 on the Delaware-Maryland border. The construction of the Interstate System was being compared to JFK's personal goal of putting a man on the moon. It was November 15, 1963.

"When was the last time the President of the United States showed up to a highway's ribbon cutting?" Bart Mitchell of Superior Paving Corp., a hot-mix producing/contracting firm in Virginia, asked in awe over 40 years later. He is able to give an eyewitness account of President John F. Kennedy's last ribbon cutting.

Asphalt was a family business for Mitchell. His first job

President John F. Kennedy cuts the ribbon opening the Delaware-Maryland Turnpike, just a week before his death in November, 1963. On left, State Senator William S. James and Baltimore Mayor Theodore R. McKeldin, JFK, Governor J. Millard Tawes and Maryland Roads Commissioner John B. Funk. Congressman George H. Fallon, chairman of the House Public Works Committee, is in glasses and hat behind Tawes.

when he graduated from college was working on the company's project in Cecil County, Maryland, an Interstate System job, running 18 miles between the Susquehanna River and the Delaware state line.

"Secret Service was out there securing the area for the ribbon cutting. I was on the site operating one of our rubber-tired rollers when they directed me to prepare a spot for Marine One to put down. The agents said, 'That's where the President will be landing, can you prepare the site for us?' There I was, running the roller back and forth just a few minutes before the President's helicopter touched down," recalled Mitchell. At the conclusion of the ceremony, Mitchell picked up the Maryland state flag that fell from the limousine carrying Maryland's Governor J. Millard Tawes, who had arrived by land to join the airborne President.

Exactly one week later, on November 22, President Kennedy was assassinated in Dallas. "Having just seen the President made his death all the more unreal," reflected Mitchell. The state flag remains a cherished keepsake of Mitchell's, reminding him of that special day JFK cut the ribbon on I-95.

BIG EVERYTHING

Regardless of the inroads concrete promoters were making on the paving of America's new Interstate System, business was good for those paving with asphalt. There was a threefold increase in asphalt tonnage produced between 1950 and 1960, going from an estimated 50 million tons in 1950 to about 150 million tons 10 years later. Between 1960 and 1970, there was another big jump, a 66 percent increase in total asphalt production.

More remarkable than total tonnage being laid was the fact that more than one third of all roads being built were getting a base constructed out of asphalt. Asphalt was being used more and more under the road as well as over it. The term black-on-black describes the process of placing a layer or layers of asphalt as base, then placing another layer of asphalt as the travel surface.

Township engineer C. W. "Duke" Beagle knew asphalt had what it would take to build a lasting superhighway. In Woodbridge, New Jersey, not far from the New Jersey Turnpike, Beagle began experimenting with, and encouraging the use of, thick black bases of about 12

Thick lifts (layers) of asphalt were pioneered in the 1950s and 1960s by Woodbridge, New Jersey, township engineer Duke Beagle. Beagle championed the technique and advised many states on its use.

AASHO ROAD TEST

"A parachute flare exploding high in the air was the dramatic signal for 60 trucks to begin running simultaneously, October 15, over five test loops of the AASHO Road Test, near Ottawa, Illinois," enthused a highway journal in 1958. Taking place from 1958–1961, the road test had 16 short-span bridges in addition to its seven miles of two-lane pavements, evenly divided between asphalt and concrete.

It was a cooperative affair. Financing came from the highway departments of all 48 states (the members of AASHO) plus the U.S. territories, the Bureau of Public Roads, and private industry. The trucks were furnished by the Department of Defense, and the drivers were troops from the U.S. Army Transportation Corps. In true military style, the drivers' meals were served in a mess tent. For two years, trucks built to carry rockets—now loaded with concrete blocks instead of

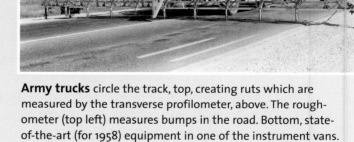

Army trucks circle the track, top, creating ruts which are measured by the transverse profilometer, above. The rough-ometer (top left) measures bumps in the road. Bottom, state-of-the-art (for 1958) equipment in one of the instrument vans.

munitions—and other heavy vehicles circled the track, aiming to destroy every bridge and mile of road.

Punishing the pavements and bridges until they failed yielded data that was quickly incorporated into pavement design guides and manuals. These research findings provided the basis for designing pavements for decades.

inches in depth. The more scientific terms for the black construction are full depth or deep strength.

As a result of Ike's Grand Plan, the public and private sectors were going gangbusters, mobilizing to the point of revolutionizing the road-building industry. Contractors were inventing ingenious machines and establishing new methods and materials, and were building more miles of superhighway faster than ever before. The 10-year period was a spectacle for those involved, one they would forever remember and never repeat. In 10 years, millions and millions of tons of earth were moved, and 23,000 miles of Interstate System were built. The Interstate Decade was the start of a half century of engineering marvels and a new heavy-equipment industry that was dedicated to superhighway construction.

Road-building gear was built in supersizes to construct the superhighways. Up until the launching of the Interstate System, small single-axle trucks with six-foot-by-six-foot cargo beds were the norm. Hauling excavated dirt from the job site, and stone and back fill to the site, these small trucks were the most common on job sites at the time. Bigger being better in the Interstate era, road-building and paving contractors' survival depended on larger hauls moving faster down the road. As a result, there was an immediate demand for heftier trucks with added hauling capacity.

Thinking bigger, contractors began using dump trucks with double axles in the rear in addition to dump wagons much larger than the new tandem dual-axle dump trucks, which were attached to the back of semitrailers. In a case of reintroduction of old technology, huge steel-bottom dump trailers hit the scene.

Cold-feed bins—
containers to hold various sizes of aggregates prior to drying and mixing—hit the Interstate on their way to installation at a new asphalt plant.

A pick-up machine and paver deal with the "windrow" of hot mix deposited by a steel-bottom belly-dump trailer.

Called belly dumps, they were remarkably similar to the form and function of their wooden horse-drawn ancestors.

Gates at the bottom of the trailer are opened as the truck moves forward, depositing a row of asphalt into the path of a "pick-up" machine that dumps it into the hopper of the paver following behind. Belly dumps were often used out west, where building the miles of highway in vast and remote areas made them popular. This type of paving is still popular in many parts of the country.

SUPER PLANTS

The old-style railroad plants were nearly extinct by the end of the 1950s. Plants in the cities were mostly on permanent sites. Breaking from the roads of iron and the traditional railroad locomotives that pulled the portable plant of yesteryear, new semitractor trucks were now roaring down the highways towing a newer and more portable asphalt plant. Easier to assemble and disassemble and capable of higher production rates, they were far more economical and practical than plants hauled over train tracks. Besides, the new Interstate Highways were being built in areas where railroads had never gone before.

In 1961, new portable asphalt plants were capable of producing 500 tons per hour. Just 10 years earlier, the 15 plants used to build the New Jersey Turnpike were each putting out about 150 tons an hour—more than a threefold improvement in a short period.

In addition to asphalt plants becoming larger and more automated, the equipment servicing the plants was improving. Arriving in the late 1940s was the front-end loader with crawler treads, making the feeding of stone into the cold-feed bins possible with the use of a single machine and resulting in a far more efficient process. By the early 1950s, rubber-tired loaders would be able to move even quicker than the durable but not as nimble tractor-treaded equipment. More mobility with rubber tires meant improved production rates—inside the confines of the asphalt plant and all along the job site.

"Are you sure about that?" "Was there ever a plant that could produce that much hot mix?" "It seems hard to believe!" challenged a group of asphalt experts in a recent and impromptu debate about the scope of past production. Hard to believe, but with an increase in demand for faster cars and bigger, wider, longer

It takes two asphalt plants, side by side, to produce enough hot mix to satisfy the demand of the contractors building a nearby Interstate.

roads came perhaps the biggest asphalt plant ever—even by today's standards. In 1973, Barber-Greene manufactured a "super plant," a mega asphalt plant that could discharge a single 21-ton batch. To put it in perspective, most plants at the time were producing between three- and four-ton batches. The super plant didn't last long, but it marked the peak of demand and production.

SURGING DEMAND

With the demand for hot mix surging, plant managers still struggled with an old problem—trucks waiting to haul off the just-produced pavement mix. The increased production levels of the Interstate Decade just made this problem more frustrating. Inventive as always, hot-mix producers partnered with equipment companies to come up with new plant equipment—first the surge bin, then the silo.

"Back in the early '60s, my dad took a trip out to California and came back with an idea for something he had seen at either Commercial Asphalt or Industrial Asphalt, I don't remember which. It was a

An early 200-ton holding silo allows the plant to keep producing while trucks are not available

surge bin—a cylindrical storage bin that could take a load of hot asphalt mix and hold it until you were ready to load it into a truck," recalled Tom Ritchie of Ritchie Paving, Inc., a Kansas-based hot-mix producer. "When he got back to Wichita, he had one built by a company that fabricated things out of steel. I was in college or maybe even high school at the time, working summers at the batch plant. It was really great. For the first time, we didn't have to stop production and all sit around and wait for the trucks." The Ritchie company's surge bin was simple— with a clamshell gate at the bottom, an open top, and no insulation. It was not suitable for storing hot mix even overnight, but it was a breakthrough.

At around the same time, hot-mix producers around the country were experimenting with the same idea. The *NAPA Newsletter* for March 1967 noted, "A two-page article on the very topical subject of surge and storage silos appears in the winter issue of 'The Spreader,' published by Warren Brothers Company. . . . The feature lists various asphalt producing companies who are active in hot-mix storage, including Teichert & Son, Inc., Sacramento, California; Dixie Asphalt Company, West Palm Beach, Florida; [and] J.D. Shotwell Company, Tacoma, Washington." A surge bin is a simpler device that can keep mix hot for only a limited time, while an insulated storage silo can keep mix ready for paving for several days. Babler Brothers, a hot-mix producer in Portland, Oregon, pioneered the drag slat conveyor, today's industry standard for moving the asphalt to the silo.

Never slow to pick up on hints from the marketplace, the asphalt-plant manufacturing companies began to introduce their own versions of silos—much more convenient for the hot-mix producer than having to design their own and get them built by local companies—and adding refinements, such as insulation and top and bottom gates that could be sealed tightly. Another essential modification was placing a batching device on the top of the silo. This helped to prevent segregation, the separation of large and small stones in the mix.

As with all new technology, there were some kinks to work out. "Our customers knew the new silos would give them an advantage in winning low-bid work on the new Interstate System," recalled asphalt industry leader Don Brock. "So, when one of our first and best customers had a problem, we jumped to get him back on line." In 1967, Brock dispatched himself and his business associate, Gail Mize, from their front office in Tennessee to get a jammed silo in North Carolina working again.

The asphalt stored in a 200-ton silo had become hardened. It was a problem that was eventually resolved with an ingenious liquid-trap-door that made an airtight seal. Meanwhile, Brock had to get his customer up and running.

"We camped out at his plant for two weeks. First we cut a hole in the side of the tank. We crawled inside with shovels, pickaxes, and jackhammers to dislodge the hardened asphalt. We parked a dump truck beneath it and threw the asphalt into that. It was miserable inside that silo!" Brock remembered.

Some skepticism about "hot storage" came from the state DOTs, the producers' biggest customers. Officials were concerned that the mix might oxidize while in the silo, or that it might segregate, with the large stones separating from the small stones. The industry worked with the agencies to address their concerns, and eventually hot storage was accepted. Large silos quickly became a familiar sight at asphalt plants. They are marketed by Asphalt Drum Mixers, Astec Industries, Custom Welding and Metal Fabricating, Dillman Equipment, Gencor Industries, Herman Grant Company, Maxam Equipment, Meeker Equipment, Reliable Asphalt Products, Stansteel/Hotmix Parts, Terex Roadbuilding, and others.

ELECTRIFYING HOT MIX

By 1960, electrical gadgetry was reliably helping asphalt plant owners increase productivity

As early as 1935, electrical devices called recording pyrometers were used at hot-mix asphalt plants to measure the temperature of ingredients. The California Division of Highways started using them a few years later and soon required them on asphalt plants providing hot mix to state road-paving jobs.

Not until the early 1950s was the first asphalt plant designed from the ground up with electrical weighing and cycling controls. By 1960, electrical gadgetry was reliably helping asphalt plant owners increase productivity by regulating the heat and composition of the hot mixes of the day.

The end of the Interstate Decade saw electronics playing a critical role in nearly every asphalt plant being built and in older ones being upgraded. By the 1970s, the use of electronics was radically improving the flow of production in asphalt plants. Additionally, electronics produced automated printouts of inventories and charted an asphalt plant's rate of production. Even bills sent to customers were beginning to be generated electronically.

BIG GREEN PAVING MACHINES

Thinking big and acting big, Barber-Greene built super-big paving machines—in some cases the biggest—painting them an identifiable green. Around the world, this all-American firm was known for a quality product and innovation. So recognized were the green machines building the long-anticipated paved roads that the name on the road gear became the designation of the road itself. In South American and African nations, any paved road might be called a Barber-Greene road. Sometimes a misinterpretation led to their being called Barbara Greene roads. "In the West Indies they would actually say, 'We're going out to Barber-Greene the roads,' instead of saying they were going out to pave," recalled industry veteran Roger Sandberg.

No matter how big Barber-Greene's machines were, they weren't immune to U.S. patent rules. When the patent on their floating screed expired after 17 years, in December 1955, their legal monopoly of its production ended. With the heady days of Barber-Greene's dominance at an end, the market was wide open. Competitors began rushing in. In the 1970s, the Barber-Greene Company lost its ascendancy. Eventually the company was sold to competitors, and the name slowly disappeared from the paving scene. Today's paver manufacturers include Caterpillar, Dynapac USA, Ingersoll-Rand, LeeBoy, Roadtec, Terex Roadbuilding, and others.

Soon several companies introduced their own floating screeds, with each company incorporating improvements and refinements. One used along the future stretches of the Kansas Turnpike had a crawler-mounted finisher with the first vibrating screed. Electronics were also introduced to the floating screed, offering automatic feeders and oscillating track rollers.

"Competition is what makes America great. When Barber-Greene's patent expired, other firms' creativity and inventiveness picked up where they left off, improving the industry," pointed out Chuck Deahl, an industry veteran. "Soon came an asphalt paver with pneumatic tires and a self-leveling mechanism, and eventually they were laying a thicker, wider lift of asphalt.

When the patent on Barber-Greene's floating screed expired after 17 years, their legal monopoly of its production ended.

"The rubber tires meant the pavers could move much faster than machines using tractor treads. That made a big difference to contractors during the early years of the Interstate System. The job sites were so darn big—they were miles long back then—driving the pavers from one end of the site to the other took a long while," recalled Deahl.

Bart Mitchell, the hot-mix producer/contractor, countered, "Rubber tires allowed gear to move faster within a job site, but tractor treads allowed gear to move better through the mud, an inherent trait on the new construction sites so prevalent during those early years of the Interstates' construction."

In the late 1960s, floating screeds capable of laying down two lanes of asphalt were introduced.

In a curious instance of technology offering two steps forward and one step backward, the floating screed allowed a faster and more consistent laydown of asphalt, but the more cumbersome and slower side-form paving machines of earlier times were capable of laying pavement twice as wide as the new, faster pavers. The older, slower machines could lay down a pavement width between 20 and 30 feet in a single pass as early as the 1920s. With all of its advances, the asphalt paver at the dawn of the Interstate Decade was limited to one travel lane of paving, about 12 feet wide. Not until the end of the 1960s was there a floating screed capable of laying down two lanes of asphalt in a single pass.

In the late 1960s, floating screeds capable of laying down two lanes of asphalt were introduced. "The new pavers were laying 24 feet of asphalt at a time. Later they could put down as much as 36 feet, the width of two travels lanes and a shoulder—the width of an entire side of an Interstate highway in a single pass," reflected Jack Farley, another asphalt veteran.

"When the Interstates were being built, this was ideal. Only later, when construction crews were going back to widen or maintain the highways, were these superwide pavers more of a hindrance than a help. The problem was traffic. When repaving with traffic using the road, crews are restricted to doing just a single lane at a time," Farley explained.

"It used to cost a fortune to lay asphalt before the floating screed. Setting the forms and getting the road to the correct grade was a lot of hard work. But laying one lane at a time was no picnic either. But with the wide screeds, contractors were moving fast and wide," said Tom Skinner, who worked for nearly a half century in the industry.

"In 1963, a 16-foot-wide asphalt paver cost $14,000. By 1970 a 36-foot-wide asphalt paver cost $60,000," Bart Mitchell recited from memory. "When those 36-foot monsters were running, it took two asphalt plants to feed them! There aren't nearly as many opportunities to use those superwide spreaders today, but while the Interstate System's construction was booming, they were ideal. Today it's mostly maintenance and repair work on the Interstates, and using 12-foot pavers is plenty on an Interstate," explained Mitchell. Pavers with wide screeds are still used today for new construction, airport runways, large parking lots, and other situations where it is desirable to pave wide widths.

Grading and leveling the superhighway's travel lanes before paving them was sometimes left to a bulldozer. Moving slowly, and dependent on an operator's skill, a bulldozer left divots from its tractor treads in the surface it was attempt-

In paving a new Interstate, two-lane-wide paving makes it fast work. Dump trucks feed material into the hoppers of two pavers, and the floating screed keeps the pavement material level so that the new pavement is smooth.

The pressure exerted by pneumatic-tired rollers can be adjusted on the fly, enabling operators to tailor the impact of the roller to the needs of the pavement. Pneumatic-tired rollers are still used in situations where extra traction is required.

ing to smooth. New types of graders came on the market, employing automation to scrape and level the earth, making for the ultimate paving conditions of the time. Using devices that acted as giant planers, they leveled the surface before paving.

The government specifications of the time sometimes called for pneumatic tires on rollers. "Air on the run" was a term describing the ability of operators to adjust the air pressure on their pneumatic-tired power rollers while staying in the driver's seat. Air could be automatically released to give a lower air pressure and soften the tires, which lessened the impact of a roller. Conversely, operators could add pressure while on the job to make for a stiffer tire to increase pressure on the pavement and achieve greater compaction.

Then came the vibratory roller. "Pneumatic-tired rollers were losing ground to hard steel-wheeled vibrating rollers, which were becoming fixtures on Interstate jobs in the mid-1960s," reflected Tom Skinner. He added, "By the 1970s, vibratory rollers were preferred over pneumatics because with their big roller drums vibrating at whatever level the operator wished, compaction was close to ideal. But they still used pneumatic-tired rollers at times." Both pneumatic and steel-wheeled rollers are still in common use. Rollers are marketed in the U.S. by BOMAG Americas, Caterpillar, Dynapac USA, Ingersoll-Rand, LeeBoy, Sakai America, Terex Roadbuilding, Vibromax America, Wirtgen America, and others.

Just as vibratory rollers were making the scene, so too were hydraulics. By the 1960s, manufacturers were installing hydrostatic

controls on power rollers, enabling them to make far smoother stops and starts when reversing the direction of the high-tonnage machines. "Everything matters when perfect smoothness is the end goal. Bringing your roller to a graceful stop and then being able to change directions without a jerking motion makes for a travel surface that's as smooth as can be," explained Allen Scott, who operates an asphalt paver for a contracting company in Maryland.

"Before hydraulics, operators had a cockpit full of levers, pulleys, and pedals to work with. Getting a large rig to cooperate meant you were kicking like hell with your feet and pulling as hard as you could with your arms. It was enough to wear out a guy in good shape," mused Jack Farley, the equipment expert. "I'll tell you this: Operating a heavy machine back in the day meant you didn't need to belong to a health club. You got your workout by the end of every day you were on the job. The combination of electronics and hydraulics has made a world of difference in the maneuverability of heavy equipment. The larger the machine, the more the difference electronics and hydraulics make."

ROUGH ROADS AHEAD

"Almost 30,600 miles of the 42,500-mile National System of Interstate and Defense Highways are now open to traffic and construction is under way on another 4853 miles, Secretary of Transportation John A. Volpe announced today," heralded a press release from the United States Department of Transportation on December 7, 1970.

John Volpe, President Nixon's Secretary of Transportation, the first Federal Highway Administrator and only the second Secretary of Transportation ever, could claim much progress. Fifteen years after his swearing-in by Eisenhower and after ground was broken, 72 percent of the Interstate System had been completed on schedule.

But road builders were calling most of the 30,600 miles open to traffic the "easy miles," meaning that they had been built in the rural areas of America without much notice or protest at the remarkable rate of over 2100 miles a year.

By the 1970s, the Interstate System's construction had moved well into the nation's downtown areas. The urban miles of Interstate System were proving to be costly and time-consuming to build. About 5000 miles, or 12 percent of the Interstate System, would be built as urban superhighway; but those miles would consume half the project's entire budget.

Work that had started with a big bang was now dragging along. Each and

Only 12 percent of the Interstate System would be built as urban superhighway; but those miles would consume half the project's entire budget.

Dressed for a more rigorous climate, a woman flagger helped expedite paving in Montana.

FLAGGERS

Women working as flaggers made news in 1962. Above, a contractor in Tennessee employed two young women in shorts as flaggers on a job site where the road was being widened from two lanes to four. The flaggers also gave out flyers suggesting alternate routes during the construction period, resulting in a 60 percent drop in traffic during working hours. A newsletter at the time reported that "the traffic that does come on through, doesn't seem to mind the delay."

every mile was now hard-fought and far more difficult to plan, design, and pave than ever before. Urbanites, community activists, and environmentalists were organizing.

There were other changes in the air. "Perhaps as a by-product of the discontent generated by opposition to the war, the decade also saw the rise of interest groups that would challenge programs whose benefits had long been unquestioned, among these the construction of highways and waterways," explained a matter-of-fact Bill Haycraft, a longtime Caterpillar Inc. executive in his book *Yellow Steel.*

Environmental consciousness was growing throughout the 1960s and 1970s. Many laws that affected highway building were enacted, including the Highway Beautification Act of 1965, the National Historic Protection Act, the Clean Air Act, the Clean Water Act, and the National Environmental Policy Act. These provisions had—and still are having—a profound effect on how highways are built. Environmentalists were closely monitoring every new construction project. Each mile of road was a battleground—or so it seemed to those trying to build highways. Regardless of feelings, construction was slowing down.

Making the complications of complying with the Highway Beautification Act look like child's play, the Clean Air Act of 1970 changed many industries, of which highway building was one. At this time, asphalt plants were not known for their environmental friendliness. Having spewed black smoke from the fuel they burned and dust from the aggregates they handled for decades before, they were now far cleaner than they had ever been, but still they were targets of the newly empowered activists.

Recognizing the importance of controlling emissions, the National Asphalt Pavement Association created guidelines for its producers in order to meet the President's Air Quality Act of 1967, a precursor to the broader act of 1970.

It wasn't enough. Air-quality issues were becoming the top agenda items at the association's national meetings in the early 1970s. Raising their standards and improving their practices again, hot-mix producers began making themselves a lot less noticeable. Odors, smoke, and noise were reduced further.

The national association encouraged asphalt-plant beautification. What at first seemed to be an oxymoron took root with greenery and putting greens. Nonetheless, negative impressions of the industry sometimes

Two secretaries of the Gallagher Asphalt Corporation try their putting skill on the company putting green during a break in their daily routine.

Working on roads already open to traffic . . . demanded smaller and smarter paving machines.

lingered in the public's mind.

As if times were not hard enough for hot-mix producers, the Yom Kippur War broke out. The events of that war triggered the oil embargo that began in October of 1973 and lasted for six months. The OPEC boycott, protesting the West's support of Israel in the war, just about quadrupled the price of a gallon of gasoline from 30 cents a gallon to as much as $1.20.

Lines at the gas pumps symbolized the lack of supply and a panic-driven demand. As the nation fell into an energy crisis, supplies of asphalt cement declined by nearly a quarter from the year before. The Dow Jones Industrial Average declined by nearly half.

Adding to the pressures on road builders were the issues of cost and the schedule of its biggest paving program. Originally, the Interstate System was scheduled for completion in 1972 at a cost of roughly $27 billion. Over the coming decades, work would drag on at the painfully slow rate of less than 500 miles a year well into the 1990s. The final total for completion, not adjusted for inflation, was more than $130 billion.

On a brighter front, electronics and hydraulics made their grand entrances to the heavy-equipment and road-building worlds during the Interstate Decade. Now the focus would be on improving these revolutionary devices.

Instead of might and magnitude, quickness and nimbleness would become the desirable traits. Working on roads already open to traffic and in the downtown areas meant smaller and tighter construction work zones that demanded smaller and smarter paving machines, some with new articulated wheels.

Instead of a grand finale for Ike's Grand Plan, work on the Interstate System continued at a slower pace for the next four decades. Boston's Central Artery—the "Big Dig," to be completed in 2006—provided the last link in the chain.

A GRAND VISION

With a half century of perspective, it's important to see how Ike's Grand Plan has changed America, explained former Massachusetts State Senator John M. Quinlan, who was a U.S. Senate staffer when Eisenhower was in the White House and knew John Volpe personally.

"As the Supreme Commander of the Allied Forces, Eisenhower achieved one of the world's greatest military victories. As President of the United States, Eisenhower launched the construction of the Interstate System, achieving one of the greatest peacetime initiatives in our country's history," said Quinlan. "Every day the Interstate System saves lives, strengthens the economy, and provides one of the most effective shields in our homeland security. All this while strengthening our economic and military positions abroad."

More objectively, *Washington Post* columnist David Broder reported on October 10, 1990, that the Interstate System was "the greatest public-works investment in the last fifty years . . ." and "has proved to be the most important economic-development strategy of the federal government."

At about the same time the *Washington Post* writer was singing Ike's praises, President George Herbert Walker Bush, whose father, a Senator from Connecticut, had backed Ike's Grand Plan from the start, signed into law the Interstate System's new name: the Dwight D. Eisenhower System of Interstate and Defense Highways.

CHAPTER FOURTEEN

Paving the Way to the Future

"Hot-mix asphalt technology has turned into rocket science in the past 10 years."

260   **AS THE INTERSTATE SYSTEM** matured, highway builders found new ways to solve old problems. One of those was how to reuse the materials from which roads are built. The idea of recycling road materials has always had its appeal. The Romans maintained their roads by removing stones and then returning them to the roadbed; McAdam also invented a system for renewing his road surfaces through reuse of the original materials. The builders of early asphalt roads in America, including the Warren Brothers, sometimes tried to reuse sheet asphalt but, with the technology available at the time, found that impracticable.

As recently as 1965, recycling asphalt pavement was an idea whose time had not yet come. But the concept interested James Gallagher of Gallagher Asphalt Corporation in Illinois—a son of the man who had provided liquid refreshment on street corners to help sell the idea of paving. He contacted one of the country's top

Milling machines developed in the 1970s are used for rehabilitation. They also make recycling easier.

asphalt research engineers, Charles Foster, director of engineering at the National Asphalt Pavement Association (NAPA), with an inquiry about "reuse of asphalt pavement."

In a letter to Gallagher, Foster wrote that he did not recall any technical articles on the topic. He added, "I made a literature search with our IBM cards on the reuse of asphalt pavement and no articles were retrieved." He enclosed a four-page list of the scholarly journals from sources such as the Highway Research Board and the Association of Asphalt Paving Technologists that did *not* include any mention of asphalt recycling.

Gallagher was one of several contractors who took an interest in recycling asphalt in the late 1960s and early 1970s. Robert Mendenhall of Las Vegas Paving Company developed a dryer for recovered asphalt, with flues that kept the flame and hot gases of the dryer from making contact with the mix. He also mixed the reclaimed pavement with a proprietary "rejuvenating agent." On December 28, 1976, he was awarded U.S. Patent No. 4,000,000 for the rejuvenating agent. Las Vegas Paving performed the first recycling job on Interstate 15. Mendenhall's contributions to recycling were noteworthy, but he was only one of many hot-mix producers and equipment manufacturers whose innovative thinking made asphalt recycling a possibility.

"THE WORLD'S LARGEST ASPHALT MINE"

Milling machines were not originally intended for recycling, but primarily were intended for maintenance.

The 1973 OPEC oil embargo stimulated thinking about ways to reduce the consumption of oil, and the 1979 oil crisis sparked by the fall of the Shah of Iran gave such efforts another boost. Aside from the fact that recycling was a magic word in the 1970s, there were other incentives: lack of supply of asphalt cement, and the skyrocketing price of anything derived from crude oil.

"Liquid asphalt in 1972 was $20 a ton all over the world. It had been at that price for more than 20 years," recalled Bill Swisher, an industry veteran. "The price peaked in 1981 at $200 a ton, a tenfold increase. The world's largest asphalt mine was the roads of the United States. There were millions of tons of asphalt cement in those roads."

Entire slabs of asphalt are sometimes removed from roads or parking lots. These slabs can be taken to an asphalt plant, stockpiled, crushed, and graded. Aside from reclaimed slabs, however, there was no equipment that worked well for "mining" the stuff. That obstacle was overcome when milling equipment was

Milling, also called cold planing, makes it possible to rehabilitate a road while keeping it open to traffic. Bottom, the pavement before milling; center, the ski controls the cut depth; below, a milled surface is smooth enough to drive on but is ready for the new asphalt overlay.

developed—although milling machines were not originally intended for recycling, but primarily were intended for maintenance. These machines have teeth on a cutting drum. The drum rotates with an upward motion, cutting the pavement up in chunks and moving the material onto conveyors that deposit it in a truck bed.

The weakness of the earlier machines was that the new surface they created was plagued with the same bumps that were present in the old surface. With the addition of a ski alongside the machine which controls the cut depth, it became possible to maintain the cutting head at a steady angle. The ski allows the action to be controlled very precisely, leaving a surface smooth enough to drive on—sometimes more drivable than the original one.

As with any new technology, there were kinks to work out. "The most expensive part was the teeth, and they didn't last long. Mining teeth were available, so that gave us a starting point. We needed to make them inexpensive and more durable. The whole industry worked on the teeth," Swisher recalled. "In fact, the whole industry worked on everything having to do with recycling." In the 1970s and 1980s, manufacturers included CMI Inc. and Ingersoll-Rand. Today, the names have changed; the machines are marketed by Caterpillar,

Night work in the Holland Tunnel, 1976. The contractor milled the pavement (foreground) prior to placing the new asphalt overlay.

Dynapac, Marini America, Roadtec, Terex Roadbuilding, Wirtgen America, and others.

Not only did the milling machine make recycling more efficient, it solved the problem of buildup of asphalt layers. Some city streets had had so many overlays, the manhole covers and curbs were inches below grade. The buildup problem was acute in tunnels, where decades of overlays had reduced clearance. In 1976, the Holland Tunnel from New Jersey to New York was the first tunnel to take advantage of the new machines to maintain clearance.

BRINGING RECYCLING HOME

Southern California hot-mix producer Dick Sprinkel heard about recycling at NAPA's 1975 annual convention in New Orleans. He recalled, "I took the idea home and talked to my brother Reed and said, 'What do you think of this?' And he said, 'Well, let's give it a try.' And so we took our 4000-pound batch plant over in Gardena and added a fifth bin to the back side of it. We had a conveyer belt coming up to that. And on that we put the RAP [reclaimed asphalt pavement], as we called it, the crushed or ground-up asphalt, and inserted that into the mix.

"Prior to starting the whole process we would test the RAP and determine how much liquid asphalt was in it. And then we would know how much we could reduce the amount of asphalt that we had to add to the new mix. So the RAP and the new material would go into the weigh hopper together, the hot material first and then the RAP on top of it. And then the liquid asphalt. And we were able to produce a very high-quality hot-mix asphalt with 10 to 15 percent of recycled materials.

"It's interesting to mention that from there we had an uphill battle, because we now had to convince [the local road authorities] that this was as good as or better than regular hot-mix asphalt. So we went out with a series of seminars and conferences with various people . . . and I would take cores and test strips to the meetings, and I'd say, here is a core of regular hot-mix asphalt and here is a core of recycled asphalt. They look the same," Sprinkel concluded.

It took more than looking at samples to prove that recycling did in fact result in a pavement that could perform as well as one made of all-new materials. The owners of the highways—the state departments of transportation (DOTs), stewards of the public trust—needed proof that performance and durability would actually be as good as in conventional pavements. Years-long research programs that included both laboratory tests and field trials were undertaken by DOTs and universities across the country. A professor at Texas A&M University, Jon Epps, led the national effort to address technical concerns with recycled asphalt. He was among the first to encourage contractors and agencies alike to treat RAP as an engineered product rather than a waste material. The prevalence of asphalt recycling is a testament to decades of partnership between the Federal Highway Administration (FHWA), the American Association of State Highway and Transportation Officials (AASHTO), state DOTs, industry, and academia.

In 1993, the FHWA and the Environmental Protection Agency (EPA) issued a report on recycling of various materials, including both consumer goods and construction materials. It showed that America's highway contractors reclaim more than 90 million tons of asphalt each year, and over 80 percent of that total is either reused in shoulders or road base or is recycled directly into hot mix.

Asphalt recycling reuses the most costly part of the pavement, the asphalt cement, in the same way that it was originally used. Other construction materials are recycled by crushing and are then used as aggregate or fill. Reclaimed asphalt pavement is a vital part of the new hot asphalt mix, not just an inert filler.

Milling machines are familiar sights today along highways and city streets, yet few Americans driving past them realize that they are witnessing a small part of the country's biggest recycling effort.

It took more than looking at samples to prove that recycling did in fact result in a pavement that could perform as well as one made of all-new materials.

DRUMS FOR HOT MIX

As production levels rose and environmental regulation took hold, hot-mix producers and the companies that manufactured equipment for them had to design new, more productive, cleaner-operating equipment. The needs of the marketplace spurred countless innovations from both contractors and equipment companies. Inevitably, some new ideas proved to be dead ends, but engineering solutions were found for the problems of the time, and technology improved to such an extent that a 1960s hot-mix producer would not recognize today's asphalt plant.

There were two major types of plants up until the 1970s—batch plants and continuous-mix plants. In a batch plant, the hot aggregates and asphalt cement would be weighed into a hopper, mixed, and loaded into a truck. In a continuous-mix plant, the stone was dried and then mixed with asphalt cement in a continuous pug mill (*see* chapter 7, "Hot Mixing, Raking, and Rolling," page 108). Of the two, the continuous-mix plant was easier to move.

In all asphalt plants, the revolving drums that are used to dry aggregates are similar to clothes dryers—but large enough to handle tons of stone at a time. Large blades or "flights" keep the rocks agitating as the drum turns. In a batch plant, the rocks come out of the dryer and are mixed with asphalt cement in batches. In a continuous-mix plant, the stone would come out of the dryer and be conveyed through a gradation unit into a pug mill. The drum-mix plant is different from these in that, after the stone has been dried, it stays in the drum, where it is mixed with the asphalt.

In the 1930s, K. E. McConnaughey pioneered a method of mixing asphalt in a drum with an emulsion. His cold-mix process, dubbed the turbulent-mass concept, was adapted by a few contractors who put burners on the drums to heat them. Experimenting with mixing in the drum in the 1960s was Tim Shearer, a hot-mix producer in Washington state.

Drum mixers began to come into their own in the 1970s. This new type of plant proved to be energy efficient and, having fewer moving parts than batch plants, less prone to expensive breakdowns and downtime. The first wave of the new generation of drum-mix plants were portable units. Soon, hot-mix producers were asking manufacturers for stationary versions of these plants, and within 10 to 15 years the drum plant was the dominant technology. Asphalt

Hot storage silos, left, are part of a 2005 asphalt plant in Utah. The baghouse is topped by its vent system, the inverted U at left center. The drum mixer is at center. Vertical and horizontal storage tanks for asphalt cement and fuel oil are at right.

plants are currently manufactured by Asphalt Drum Mixers Inc., Astec Industries, Dillman Equipment, Gencor Industries, Terex Roadbuilding, and others.

When the Clean Air Act of 1970 became law, many industries, including the hot-mix asphalt (HMA) industry, were confronted for the very first time with a real necessity to reduce their impact on the environment. The newly created EPA began to set standards for allowable levels of emissions, but when the standards for HMA plants were first published, the technology for meeting those standards did not yet exist, was prohibitively expensive, and/or was difficult to obtain.

At a hot-mix plant, the hot air and water vapor vented from the dryer contain powdery particles of rock known as fines. The challenge is to remove the fines from the gases before they are released into the atmosphere. The first solution that became popular was the wet scrubber, but ultimately this technology was not fated to prevail.

Soon the asphalt industry turned to baghouses, standard industrial hygiene equipment that was already being used in many industries. A baghouse is like a gigantic shop vacuum, in which fines and dust are collected on the outside of bags, while clean air passes through the center of the bags and is released. The fines can then be returned to the mix or stored for later use.

Adapting baghouses to asphalt plants required learning how to design and size them appropriately. Although the technology was expensive, hot-mix producers soon found that baghouses were more effective than other control equipment and actually had a payback by using less horsepower, requiring less fan maintenance, and providing the ability to reuse the fines in most cases. At first, HMA producers bought their baghouses from the companies already in the business. Today baghouses designed specifically for hot-mix plants are made by Asphalt Drum Mixers Inc., Astec Industries, Custom Welding and Metal Fabricating, Dillman Equipment, Gencor Industries, Maxam Equipment, Meeker Equipment, Reliable Ashalt Products, Stansteel/Hotmix Parts, Herman Grant Company, Terex Roadbuilding, and others.

Vacuum cleaner for air. Dust-laden air enters the baghouse and is sucked onto the bags, trapping the dust particles, called fines, on the outside of the bags. Periodically, a strong burst of air goes into the bags, forcing the fines to drop to the floor, where they are collected and metered back into the hot mix. Clean air is vented out the top.

THE $10-MILLION QUESTION

While recycling technology was gathering steam across the country, the asphalt industry was beginning to get a black eye because asphalt mixes of the time were not holding up to the forces exerted on them by the larger, heavier trucks that were now traveling the nation's highways.

Without a doubt, changes were needed. The asphalt producers in the 1980s wanted the benefits of comprehensive programs of asphalt research to provide answers to these questions and others. And, they wanted the research to be highly practical and useful to contractors.

The members of NAPA, led by President John Gray, took on the challenge, deciding to raise $10 million to endow a research center through the NAPA Education Foundation. Two producer/contractors, Ronald Kenyon of Iowa and Harry Ratrie of Maryland, spearheaded the fund-raising effort. The entire industry joined in supporting the center.

A key part of the vision was to make the new institution independent of the for-profit companies in the asphalt industry by finding it a home at a university and creating a governance structure that would give a say in its running to both researchers from academia and officials of federal and state highway agencies. The aim was not to build "NAPA's lab" but to create a research institution whose integrity and credibility would be beyond dispute.

NCAT was a hot topic for the hot-mix industry in 1988. Discussing the new facility at an open house in Auburn are Harry Ratrie, NCAT chairman; Ronald D. Kenyon, NAPA chairman; Thomas B. Deen, executive director of the Transportation Research Board; and Frank L. "Chip" Whitcomb, chairman of the NAPA Education Foundation.

Impatient to get started, the NAPA board of directors decided not to wait until the entire $10 million had been raised. In 1986, after a nationwide search and with $5 million in the kitty, the NAPA Education Foundation signed an agreement with Auburn University in Alabama, making it the home of the National Center for Asphalt Technology (NCAT). The remainder of the $10 million endowment was raised within a few years.

In 1986, it seemed daring just to plan to fill up the 5000-square-foot lab with equipment. Less than 15 years later, NCAT opened a 1.7-mile test track and a 40,000-square-foot research center, complete with research labs, teaching labs, offices, and classrooms.

SHRP FOCUS ON HMA

At around the same time that NCAT was being created, the federal government's ambitious Strategic Highway Research Program—known by its letters as "SHRP" or simply pronounced "sharp"—was getting under way. From 1988 to 1993, this $150-million effort, funded by Congress through the FHWA and administered by the Transportation Research Board (TRB), addressed the effect on pavements of the staggering, unanticipated increases in traffic loadings that pavements had

endured in the previous 10 to 20 years. Almost $50 million of the SHRP money was spent on asphalt research.

At the beginning of the SHRP process, some in the industry thought it would result in a "more forgiving binder." As research proceeded, it was concluded that both the binder and the aggregates needed to change, and that the system for designing mixes needed work, too. The result was a new mix design system called Superpave, which stands for SUperior PERforming Asphalt PAVEments. Superpave provided the first radical departure in mix design for America's pavements since the Marshall method was popularized during World War II (*see* chapter 11, "Paving the Way to Victory," page 182).

Superpave binders are specifically chosen according to local temperature extremes and traffic conditions. The mixes are also adapted to traffic levels. Depending upon the placement in the pavement structure and the type of road being constructed, initial costs can be higher than for conventional mixes, as Superpave sometimes uses polymer-modified binders and higher-quality aggregate; but the presumably longer life offered by these pavements yields a lower long-term cost. The first Superpave pavements were placed in the early 1990s and soon many states were using the new system.

Superpave provided the first radical departure in mix design for America's pavements since the Marshall method.

AMERICANIZING *SPLITTMASTIXASPHALT*

In the never-ending search for longer-lasting pavements, word reached the U.S. from Germany of a new premium surfacing, *Splittmastixasphalt*—translated into stone-matrix asphalt, or called SMA for short. Compared to a tradition-al dense-graded mix, the idea of SMA was as foreign as its origins. It used a "gap-graded" design, meaning that there were some small stones and some large ones, but few medium-sized ones. In addition, cellulose or other fibers were added.

After reading technical papers on the subject, a joint agency-industry-university group arranged a study tour to Europe in 1990. Contractors joined researchers and representatives of both the FHWA and the state DOTs in observing both plant production and laydown operations.

The tour was a resounding success, and so was SMA. Officials from the Michigan and Maryland DOTs started plans for their first SMA pavements as soon as they returned. Maryland and Georgia quickly embraced SMA as the surface of choice for high-volume highways. As success in these two states became widely known, Colorado, Illinois, Indiana, Louisiana, Mississippi, Missouri, Virginia, Wisconsin, and other states also adopted SMA enthusiastic-ally. Other states joined the trend, and SMA became popular throughout the country for its durability, smoothness, and noise-reducing properties.

SMA and other premium mixes take advantage of stone-on-stone contact to provide a strong skeleton for the pavement. The shape of the stones is very important in avoiding rutting. Asphalt mixes containing angular stones are more resistant to rutting and shoving than mixes with round ones. To see the principle at work, fill a bowl with marbles and another with dice. Then make a fist and try to drive your knuckles into each bowl. It will be easy with the marbles—being smooth and round, they will quickly roll away from the pressure point. The dice, though, will form interlocking layers.

Stone-matrix asphalt is a tough, rut-resistant pavement which uses hard, cubical, durable aggregates. Below, squared-off rocks illustrate how the stones used in SMA interlock to form a strong skeleton.

"THE RIGHT THING TO DO"

Many industries have reacted to regulation by resisting it. The hot-mix industry decided to embrace a different approach: partnering. Together with government agencies and the unions that represent the workers, hot-mix asphalt producers set about improving conditions in the workplace and reducing the impact of HMA operations on the environment.

Over the years, the hot-mix industry has formed partnerships designed to protect the environment and to enhance workers' health and safety. It has cooperated not only with the EPA but also with the National Institute for Occupational Safety and Health, the Occupational Safety and Health Administration, the Center to Protect Workers' Rights, the Laborers' International Union of North America, and the International Union of Operating Engineers to address these concerns.

Don King, a hot-mix contractor from New York, recalled, "We realized that partnering was the right thing to do from a moral standpoint and from a practical standpoint. We have all seen industries that resist regulation, and the result is regulators who come in and fine you and hold a hammer over your head. Instead of that, we were in favor of working with the regulators prior to regulation. We felt that a cooperative effort would help our industry improve its emissions."

In 1989, the industry initiated a program for testing air emissions at HMA plants in partnership with the EPA. "The initial question was, what in fact do we put out of our stacks?" according to Don King. "We felt certain that HMA plants were not major polluters, but there was not enough data to back up our thinking. We felt we had a good case, but at the time we really couldn't prove it."

In 1989, the industry initiated a program for testing air emissions at HMA plants in partnership with the EPA.

NAPA and the U.S. EPA sponsored testing at several HMA facilities at the time and agreed on a test protocol. The goal of the program was to quantify emissions from HMA plants. King recalled the early meetings with EPA as "very open and frank." He added, "Writing and agreeing on the test protocol was important. We knew we did not want five testing efforts done five different ways. Before we started, everyone including the EPA was in agreement that the testing would be done to a certain protocol and would be meaningful."

In 1992, in the midst of the research program, another branch of EPA the published a list of sources—including HMA plants—targeted for MACT (Maximum Achievable Control Technology) standards, as directed by the Clean Air Act of 1990.

It took some time to work through all the data. By 1994, however, it was time to approach the EPA's Office of Air Quality, Planning, and Standards, located in North Carolina. NAPA requested that, based on the data, HMA plants be removed from the list. Once again NAPA and the EPA sponsored extensive testing at two hot-mix plants to fill gaps in the data base.

The EPA solicited additional data from various state air-pollution organizations and incorporated up to 400 additional test reports. Through this process, it was determined that the EPA/NAPA data represented the greatest knowledge resource available about emissions for HMA facilities.

Plant testing in North Carolina, 1997, was one of the cornerstones of EPA's research into air quality at hot-mix asphalt plants. Additional tests were conducted in Massachusetts and California, on both batch and drum-mix plants. The testing program produced more than 6000 pages of data. Few, if any, industries have been evaluated as thoroughly as the hot-mix industry by the EPA.

Up to this point, both the asphalt pavement industry and the air-quality experts at the EPA had focused their attention on emissions from the stacks at asphalt plants. In 1995, however, a group of concerned citizens in Massachusetts raised questions about "fugitive" emissions from the top of silos and at the point of truck load-out. NAPA and EPA worked with the citizens' group to put together a new testing program focusing on blue-smoke emissions from those areas. All the partners wrote the protocols together and were on-site to observe the process of testing. This 1996–1998 testing program resulted in 6000 pages of test results. It is estimated to have cost the EPA in excess of $1.5 million. Few, if any, industries have been evaluated as thoroughly as the HMA industry.

Once deliberation was completed, the EPA issued its ruling. In a notice published in the *Federal Register* on February 12, 2002, HMA production facilities were removed from consideration for MACT standards. In November of 2002, the EPA published a second *Federal Register* notice that also removed HMA production facilities from a list of industries that contribute to 90 percent of nationwide emissions of seven airborne substances that are known to be particularly harmful. In the November notice, EPA used the words "trivial" and "negligible" in describing why sources such as HMA production were being removed from their list.

Being "delisted" by the EPA as a major source of hazardous air pollutants (HAP) is quite rare. Out of 174 source categories targeted for the stringent new standards over the past 10 years, the hot-mix asphalt industry is one of a very few that have been delisted.

Thanks to many years of development of new technologies and processes, emissions from today's asphalt plants are very low and well controlled. Remarkably, total emissions from asphalt plants declined by 97 percent between 1960 and 2005—despite a 250 percent increase in production.

The testing program led to the deletion of asphalt plants from EPA's list of industries that are major polluters.

TESTING THE TESTS

Hurricane Georges dumped 5 inches of rain on Auburn, Alabama, on September 29, 1998. Despite the weather, the contractors and dignitaries gathered from across the country proceeded with their task: breaking ground for a new pavement test track at NCAT, the asphalt research facility.

The 1.7-mile track, located on 309 acres, received support from the FHWA, Alabama DOT, and several other state DOTs. It was designed to test surface mixes from several states. But its mission was greater than that: It was also designed to

A soggy start for the new pavement test track. Breaking ground are Bill Muse, Bill Walker, Paul Parks, John Spangler, Ray Bass, Mike McCartney, Don Gallagher, Dale Decker, Tim Docter, and Ray Brown.

test the tests. In spite of the millions of dollars and many years invested in asphalt research, the question of which lab tests could be counted on to predict the performance of particular asphalt mixes under actual road conditions remained unanswered.

One of the most realistic and accurate testing approaches is to use a technique called accelerated loading.

A well-known accelerated-loading test was performed in the mid-1990s at Westrack, near Reno, Nevada, where the then-new mix design process known as Superpave received some of its earliest field trials. Westrack was unique in that it used driverless trucks—computer-controlled semitrailers.

Westrack's trucks—minus the driverless technology—found a new home at NCAT in 2000, where, in the first round of tests, four fully loaded trucks, weighing 152,000 pounds each, circled the track and gave the pavement 10 to 12 years' worth of punishment in just two years. The trucks were driven over 1.6 million miles at 45 mph in that time. If an individual drove 30,000 miles a year, it would take over 50 years to drive that far.

The pavement base was thick, nearly two feet. What was being tested was the surface mixes. Nine states sent their own locally available asphalt cement binders and aggregates, and the mixes were made according to the states' own specifications. Forty-six test sections were constructed. Moisture gauges and temperature probes were implanted in the pavement sections.

The researchers were looking for rutting; they wanted to compare the results on the track with the results of the tests that had been performed on the pavements in the lab prior to construction. When the first phase of testing ended in fall 2003, the research findings were impressive: The worst-performing sections had only a quarter inch of rutting, and the average rut depth was one eighth of an inch. The deepest ruts were in the mixes that were "designed to fail," but even those ruts were so small that they made it difficult for the researchers to draw conclusions.

Clearly, the research that had been conducted since the mid-1980s had paid off. Within the asphalt industry, research is a cooperative venture. Much research

A heavily loaded truck on the oval test track in Alabama. The drivers say boredom is not a problem, and they like driving the big rigs while staying close to home. Inset, track manager Buzz Powell and NCAT director Ray Brown view test results, which are posted on a Web site at www.pavetrack.com that is updated continuously.

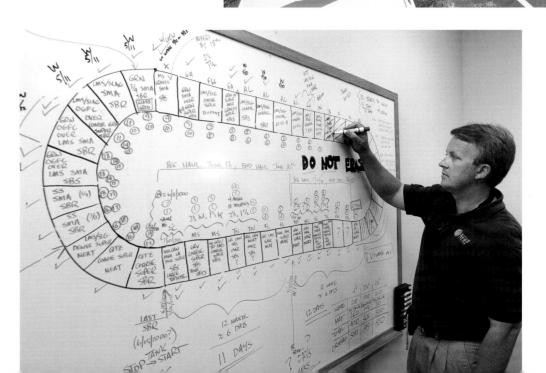

Powell fills in the details about one of the 46 test sections, left. Sponsoring state DOTs—Alabama, Florida, Georgia, Indiana, Mississippi, North Carolina, Oklahoma, South Carolina, and Tennessee—shipped in their own aggregates and asphalt cements, and the sections were built to each state's specifications. FHWA also sponsored sections.

Resonant pavement breaker has a breaking shoe that uses low-impact vibrations to reduce the concrete pavement to rubble—good rubble. By leaving the rubblized pavement in place to form the base for the new asphalt road, many trips in haul trucks are saved; congestion at the paving site is reduced; less diesel fuel is consumed by haul trucks; and the amount of stone quarried for the pavement is reduced.

Multiple-head breaker repeatedly drops 1000-pound hammers to the surface, breaking the pavement into pieces. This prevents "reflection cracking," in which the cracks in the concrete eventually reflect up through the overlay and make it rough. Rubblization is very cost-effective when compared to complete removal and replacement of the existing pavement.

is conducted at universities. The FHWA's Turner-Fairbank Highway Research Center in McLean, Virginia, is also a state-of-the-art institution that does important work. The FHWA, AASHTO, and TRB all play vital roles in conducting or sponsoring research and disseminating the results.

Thanks to all these efforts, materials and practices in the industry had changed drastically, and some of the pavement problems that had seemed intractable in 1988 were nonissues by 2000.

Once the first round of testing was completed, round two began. Some pavement sections were left in place for further testing. Others were removed and replaced with mixes that would test various structural designs. The heavy trucks are still circling the track in Alabama, helping researchers find better ways of building asphalt pavements. As this book goes to press at the end of 2005, preparations are being made for round three.

MAKING OLD PAVEMENTS NEW AGAIN

In 2001, approximately 69 percent of Arkansas's 35-year-old concrete Interstates were in poor condition. Patching and spot repairs were no longer enough, but concrete pavements that have reached the end of their useful life have to be completely removed and reconstructed. If this could have been done overnight, and at little expense, it would have been like Cinderella's transformation on the night of the ball.

Lacking a fairy godmother, the Arkansas State Highway and Transportation Department (AHTD) came up with a real-world solution that gave the state practically a new system of Interstates, with more than 300 miles of its 636-mile system completely rehabilitated and turned into modern asphalt pavements in just six years. The key was rubblization and overlay.

Rubblization and its sister technology, cracking and seating, use special equipment to break deteriorating Portland cement concrete (PCC) pavement into chunks, leaving it in place. The old pavement becomes the base for a new hot-mix asphalt (HMA) pavement. The process is environmentally friendly, because it reuses all the old PCC, avoiding waste. And it is both quick and cost effective.

In cracking and seating, a giant machine repeatedly drops guillotine-style breakers on the concrete, cracking the pavement at two- to three-foot intervals. A roller is then used to "seat" the broken pavement and make it more stable as a base. This technique, which was pioneered in the 1970s, is still used in many states.

Rubblization and its sister technology, cracking and seating, use special equipment to break deteriorating Portland cement concrete (PCC) pavement into chunks, leaving it in place.

Rubblization represents a refinement in which the surface of the pavement is broken into smaller pieces either by a resonant breaker, which uses vibratory action, or by a multiple-head breaker, which pounds the pavement with heavy hammers. No matter which equipment is used, the broken-up pavement is then compacted with a vibratory roller.

Breaking up the concrete prevents cracks in the concrete underneath from being "reflected" into the riding surface above. When this is not done, although there may be nothing wrong with the asphalt overlay itself, the underlying problems in the concrete can translate into a rough ride.

"Rubblization makes the base into an interlocked matrix of pieces as the concrete breaks up. It functions much like a jigsaw puzzle with broken pieces fitting together," explained AHTD chief engineer Bob Walters.

Although no other state has rubblized and overlaid its failed pavements on such a scale, the technique has been used extensively from coast to coast since the late 1980s. Highways, city streets, and airports have been rehabilitated in this quick, efficient, cost-effective manner. Rubblized or cracked-and-seated concrete can also form the base for Perpetual Pavements.

Perpetual Pavements have been built for decades. They have gone by names such as full-depth and deep strength.

(3) SMA, OGFC or SUPERPAVE
(2) High Modulus Rut Resistant Asphalt
(1) Flexible Fatigue Resistant Asphalt
Pavement Foundation

THE GREAT UNNOTICED MATERIAL GETS NOTICED

Perpetual Pavements have been built for decades. They have gone by names such as full-depth and deep-strength, meaning that they had a hot-mix asphalt base and a hot-mix asphalt surface. Rehabilitation was mostly a matter of milling off the top few inches for recycling, then putting down a new surface. The pavements were not prone to significant problems such as bottom-up cracking or deep structural rutting that might have made it necessary to remove the entire pavement structure and rebuild it.

Asphalt's ease of rehabilitation makes it the great unnoticed material. Although it's everywhere, people take it for granted. For asphalt contractors and highway agencies, this is a good thing; it means that the roads are doing their job, not bringing attention to themselves by falling apart and needing to be rebuilt.

In the late 1990s, road engineers from around the world began to compare their research, and several of them found that the same things were being observed in the U.S. and Europe. They began to discuss "long-life asphalt pavements." Eventually, someone coined the term Perpetual Pavement and it stuck.

Research was undertaken at NCAT, the University of Illinois, and the Asphalt Institute to determine, first, why these pavements lasted as they did; and second, how to develop guidelines for designing them consistently. Since that time, other universities and research institutions have begun their own studies of long-life asphalt pavements, and even more research is on the horizon.

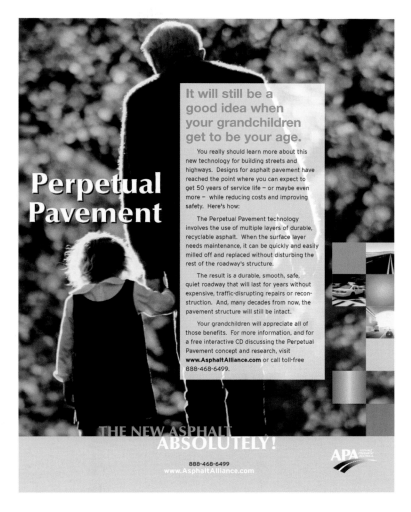

Asphalt is often referred to as a "flexible" pavement, distinguishing it from "rigid" concrete. The flexibility and resilience of the material is one key for Perpetual Pavement. It is designed so that the bottom of the pavement never has to bend so much that it becomes fatigued and forms cracks. Bottom-up cracking would lead to catastrophic failure and the need to extensively patch or totally reconstruct the pavement. This type of failure is mostly seen in thinner pavements that were not designed to withstand the weight of the traffic they actually carry.

In a Perpetual Pavement, the upper layers take all the punishment—with the asphalt being put under pressure, then released, then put under pressure again—and the bottom layer never reaches its endurance limit.

Superstition Freeway's asphalt overlay sparked citizens' demands to resurface the existing freeways in Phoenix. Inset, open-graded pavements also drain water from the surface during storms, reducing splash and spray that can obscure a driver's vision.

BELIEVERS ALONG SUPERSTITION FREEWAY

Dan Galvin winced as he scrolled down a long list of e-mails. As community-relations liaison for Granite Construction Co., Galvin had been brought in to take the heat if construction along Superstition Freeway (U.S. 60) between Tempe and Mesa, Arizona, made drivers angry. The 2002 project included widening 12 miles of road in both directions and then resurfacing with an asphalt rubber overlay that was open-graded—that is, with pores in the surface.

Galvin clicked on the first e-mail. "I just had to write and say that the new rubberized asphalt on US 60 totally rocks. . . . What is the chance of re-surfacing ALL of our highways with that stuff?" Surprised, Galvin clicked on the second message: "Congratulations to all of you on this super job, completed early! And the quiet pavement is fabulous—it should be used everywhere." And the next: "Love that new surface . . . will you be doing the off-ramps like that too?" What was so

tough about community relations, thought an elated Galvin.

Overall, the public reaction to the U.S. 60 job was massive and positive. Within weeks, e-mails to Galvin had given way to letters to the editor and calls to radio talk shows. Drivers liked the quiet ride. People living near the roadway liked not hearing the cars. Everyone else in the Valley of the Sun wanted neighboring freeways like this one—quiet.

While Arizona has used asphalt rubber pavements for years, Superstition Freeway was the first to set off a public movement. Arizona's state and local governments soon responded with a $34-million resurfacing plan for 115 miles of existing concrete freeways in Maricopa County, which surrounds Phoenix, the state capital. Even the new concrete freeways in the area are now built with a sound-absorbing surface of asphalt.

Not just in Phoenix but all over the country, quiet pavement technology is a hot topic. Building a quiet pavement makes sense; unlike a sound wall, which only reduces noise pollution for people in its sound shadow, a quiet pavement stops noise at its source. Some quiet pavements, like Phoenix's, are open-graded asphalt rubber, with the surface pores absorbing the sound produced by the tires rolling over the asphalt. SMA also provides a very quiet surface.

COOLING DOWN THE MIX

The next big thing in *hot*-mix asphalt may be making it a *warm* mix. Several new technologies that are being tested allow production and laydown temperatures to be lowered by as much as 80 to 100 degrees Fahrenheit. (Most asphalt mixes in the U.S. today are produced at 300 to 325 degrees Fahrenheit.)

Besides lowering greenhouse-gas emissions and reducing energy consumption, warm mix may cause less wear on the components of asphalt plants and paving equipment. It is even possible that in the long run, pavements may be less prone to cracking because the binder does not "age" as much during construction.

In the summer of 2002, a delegation of NAPA staff and members toured Europe, where the first warm-mix technologies originated, to see it being produced and placed on roads. They came back so impressed by what they had seen that they started a program of research to investigate whether the processes and products are compatible with the mix designs, equipment, climate conditions, and work practices in the United States, which are quite different from those in Europe.

Besides lowering greenhouse gas emissions and reducing energy consumption, warm mix may cause less wear on the components of asphalt plants and paving equipment.

"If we can make this technology work in the U.S., we can reduce energy costs, and we will also cut the amount of CO_2 we put in the air. Warm mix has benefit-benefit-benefit," said tour participant David Carlson, a hot-mix producer from Iowa.

THE NEXT 50 YEARS

"The volume of freight moved in this country is expected to double over the next 20 years, with the volume carried by trucks increasing by more than 80 percent."

"It is difficult for many to comprehend that today 3.4 million trucks, traveling over 200 billion miles a year, carry 72 percent of America's freight traffic on the nation's highways. Since 1973, annual miles traveled by trucks have increased 225 percent, but highway capacity has increased only 6 percent." Charles F. Potts, of Heritage Constructuction and Materials, cited FHWA statistics in the address he delivered when he was inaugurated as chairman of NAPA in 2004.

He continued, "Obviously, this has led to serious congestion on our nation's highways. Many of our roads are carrying more vehicles than they were designed to handle. Sophisticated new approaches in business, such as 'just-in-time' manufacturing and delivery, mean smaller and more frequent shipments. . . . The volume of freight moved in this country is expected to double over the next 20 years, with the volume carried by trucks increasing by more than 80 percent."

Quantum leaps in the amount and weight of traffic are not the only changes on the horizon. The public desires smooth, durable, safe, quiet pavements, and the public does not like congestion. Maintaining focus on the ultimate customer—the motorist—will be a challenge for both the construction industry and highway agencies. As Dewitt Greer, a legendary highway man, said, "This Nation does not have superhighways because she is rich, she is rich because she had the vision to build such highways."

In the past few years, the price and availability of both fuel and construction materials has been affected repeatedly by unexpected events—terrorism, wars, natural disasters. Such events test the ingenuity and resolve of the highway community, and the highway community has faced up to these challenges. Maintaining our mobility is of paramount importance to the nation's continued economic growth and health.

What we have seen in catastrophes such as Hurricane Katrina and

the September 11 attacks is that relief and recovery efforts depend on our system of roads. When the chips are down, Americans take to the road, first to evacuate threatened areas, next to rush to the relief of survivors, then to build anew.

While we do not know what is around the next bend in the road, we can be certain that the people who pave America's roads will continue to look for new and better ways to meet the country's needs.

Sources

CHAPTER ONE
AMERICAN ASPHALT

10 "As new and greater road-systems . . .": Frank Lloyd Wright, as quoted by Tom Lewis, *Divided Highways*, Viking Penguin, New York, 1997, p. 239.

12 more asphalt is recycled than paper, glass, metal, or wood: "A Study of the Use of Recycled Paving Material," a joint report to Congress from the FHWA (FHWA-RD-93-147) and the EPA (EPA/600/4-93/095), June 1993.

12 mountain of recycled material 749 feet high and 1398 feet in diameter: Kent Hansen, NAPA Director of Engineering, as calculated from the FHWA/EPA report to Congress, June 1993.

12 in 2002, the EPA deleted asphalt from list of major sources of air pollutants: *Federal Register*, February 12, 2002

12 National Park Service's stated quest to "lie lightly on the land": www.nps.gov/

CHAPTER TWO
ROMAN ROADS

20 "In the beginning was the road.": Joseph Bedier, quoted by Christy Borth, *Mankind on the Move*, Automotive Safety Foundation, Washington, D.C., 1969.

20 Along its 50,000 miles of highways and 250,000 miles of secondary roads: Herman Schreiber, *The History of Roads* (Translated by Stewart Thompson from *Sinfonie der Strasse*), Barrie and Rockliff, London, 1961, p. 115.

22 to split up their own fighting forces, conquering two different enemies on two different fronts: Schreiber, *The History of Roads*, p. 150.

22 Roman army legions, 5000 soldiers strong—the size of Allied infantry regiments during the Normandy invasion on D-Day: Stephen E. Ambrose, *D-Day: June 6, 1944*, Touchstone, New York, 1994, introduction.

22 Julius Caesar slept in his chariot: Schreiber, *The History of Roads*, p. 134.

22 post houses . . . every six miles along the main highway: Cyril Hughes Hartmann, *The Story of the Roads*, George Routledge & Sons Ltd., London, 1927, p. 6.

22 Contributing money to a road fund in ancient Rome: Schreiber, *The History of Roads*, p. 119.

22 up to 16 years for a typical infantryman: J. M. Roberts, *New History of the World*, Oxford University Press, New York, 2003, p. 232.

23 Sometimes wooden and stone rollers were used to compress the ground: Schreiber, *The History of Roads*, p. 119.

25 Roman engineers were the first to build road tunnels through rock: Schreiber, *The History of Roads*, pp. 126, 127.

CHAPTER THREE
MACADAMIZING

27 "But who effected this improvement in your paving?" . . .: Roy Devereux, *John Loudon McAdam*, Oxford University Press, London, 1936.

27 "Few people seem to realize what McAdam did for this country. . . .": Devereux, *John Loudon McAdam*, p. 1.

28 3000 coaches, 150,000 horses, and 30,000 coachmen . . .: Deveraux, *John Loudon McAdam*, p. 56.

28 livestock, cheeses . . . tin and lead: Devereux, *John Loudon McAdam*, p. 3.

30 "It appears that Mr. McAdam . . .": John Loudon McAdam, Esq., *The Present System of Road Making (McAdam on Road Making) With Observations Deduced from Practice and Experience*, 1823, 7th Ed., Longman, Hurst, Rees, Orme & Brown, Paternoster Row, p. 11.

31 Telford's roadway designs more elaborate than Trésaguet's and thicker than McAdam's . . . large stones: Billy Joe Peyton, *The National Road* (Chapter: "Surveying and Building the National Road"), Johns Hopkins University Press, Baltimore, 1996, p. 141.

31 stones of six ounces or less: Devereux, *John Loudon McAdam*, p. 3.

31 travel surface of road called the roof: Devereux, *John Loudon McAdam*, p. 52.

32 On October 8, 1824 . . .: Devereux, *John Loudon McAdam*, p. 64.

33 Jackson as "violent, quarrelsome, vigorous, brusque, and uncouth"; fought British at 13: E. J. Applewhite, *Washington Itself*, Alfred Knopf, 1986, p. 83.

34 constructed with heavy stone foundations: Peyton, *The National Road*, p. 141.

35 maintenance and repair budget of $88 a mile: Peyton, *The National Road*, p. 149

35 macadamizing a 73-mile section: Peyton, *The National Road*, p. 150.

35 old surface recycled into a macadamized roadway: Peyton, *The National Road*, p. 143.

CHAPTER 4
ASPHALTING THE AVENUES

38 *I stop the noise from the street, so appalling . . .:* "Asphalt's Soliloquy" by W.S. Godwin, of the Texas Co., Hugh T. Boorman, *Asphalts, Their Sources and Utilization*, William T. Comstock, New York, 1908

38 hickory brooms, necklaces, and men riding horses clad in bark: Mary Cable, *The Avenue of the Presidents*, Houghton Mifflin Company, 1969, p. 66.

38 pigs, chickens, and grazing cows: Frank L. Dyer, "The Streets and Roads of Washington," *Good Roads*, June, 1892, pp. 316, 318, 322.

38 "soon to be beautiful": Carol M. Highsmith, Ted Landphair, *Pennsylvania Avenue*, American Institute of Architects Press, Washington, D.C., 1988, p. 64.

38 "in rainy weather . . .": Dyer, "The Streets and Roads of Washington," p. 322.

38 "I have never seen . . .": Cable, *The Avenue of the Presidents*, p. 65.

40 to move [the capital] farther west: Cable, *The Avenue of the Presidents*, p. 145.

40 "city of streets without houses": Dyer, "The Streets and Roads of Washington," p. 319.

41 six horses to pull howitzer: Dyer, "The Streets and Roads of Washington," p. 320.

41 $6.25-million plan: Highsmith and Landphair, *Pennsylvania Avenue*, p. 64.

41 **200 miles of streets:** Dyer, "The Streets and Roads of Washington," p. 323.

41 **sewers, gas and water lines, 50,000 trees . . . 200 miles of streets:** Cable, *The Avenue of the Presidents*, p. 149.

41 **called for paving 93 miles . . . and 28 miles with asphalt:** Dyer, "The Streets and Roads of Washington," p. 323.

41 **"lifted Washington out of the mud":** Dyer, "The Streets and Roads of Washington," pp. 322, 323.

42 **Market Street as "dirtiest thoroughfare in the city":** Hugh M. Gillespie, *A Century of Progress,* National Asphalt Pavement Association, Maryland, 1992, p. 40, as reported in the *American Contract Journal* in 1882.

43 **Berlin streets; "with extreme strictness":** Leon Malo, *Asphalt Roadway,* E&F N. Spon, London, 1886, pp. 11–13, 23.

43 **Cortlandt Street paved with iron . . . replaced with stone:** *Asphalt Pavements,* Barber Asphalt Company, 1889, p. 2.

43 **cobblestones—dimensions, relegated to lesser uses:** Harwood Frost, *The Art of Road Making,* Engineering News Publishing Co., London, 1910, p. 314.

44 **Belgium blocks, description and problems, last more than 15 years:** Frost, *The Art of Road Making,* pp. 314, 317, 321.

44 **cobblestones on Pennsylvania Avenue:** Cable, *The Avenue of the Presidents,* p. 144.

44 **Paris . . . 42,000 vehicles and horses:** Malo, *Asphalt Roadway,* p. 2.

44–46 **clay bricks—"stiff mud," flattened, heat raised for a week to 1500–2300 degrees, used in Charleston:** Frost, *The Art of Road Making,* pp. 323, 325, 326.

46 **wood blocks in 1830s:** Frost, *The Art of Road Making,* p. 336.

47 **1848 patent of the Nicholson, types of wood, 240,00 per day, specifications of blocks, 22 pounds of oil:** Frost, *The Art of Road Making,* pp. 336–338, 347, 351.

48 **use of asphalt in Paris, Berlin, London:** Malo, *Asphalt Roadway,* p. 6.

48 **1802, asphalt for sidewalks, bridges, floors:** J. E. Pennybacker, "Asphalt: Nature's Most Versatile Product," Asphalt Institute, New York, Ninth International Petroleum Exhibition, 1936, p. 9.

48 **rock asphalt fell onto macadam:** Frost, *The Art of Road Making,* p. 359.

48 **summer heat creates macadam surface in Switzerland:** Amzi Lorenzo Barber, *The Best Road and the Right Way to Make It,* A. L. Barber Asphalt Company, New York, 1909, p. 6.

49 **1888 asphalt pavement in Paris and London:** "Asphalt Pavements," p. 6.

49 **asphalt only in coastal cities of U.S.:** W. F. Pollard, "Origin and Development of Rock Asphalt Deposits," North Carolina Highway Bulletin, July 1921, p. 9.

50 **1838 asphalt use at Philadelphia Merchants Exchange:** Pennybacker, "Asphalt: Nature's Most Versatile Product," p. 9.

50 **quotes about Fifth Avenue dust:** *The New York Times,* October 9, 1869, p. 1.

51 **"In the history of road building . . . ":** Barber, *The Best Road and the Right Way to Make It,* p. 4.

51 **DeSmedt added culm and sand to asphalt; attempted to make wooden blocks last longer:** Barber, *The Best Road and the Right Way to Make It,* p. 8.

51 **asphalt at Newark's city hall:** Pennybacker, "Asphalt: Nature's Most Versatile Product," p. 9.

52 **heated lake asphalt and added sand-powdered limestone and petroleum:** Pollard, "Origin and Development of Rock Asphalt Deposits," p. 9.

52 **DeSmedt's patents, inspiration for large-scale operation:** Barber, *The Best Road and the Right Way to Make It,* pp. 7, 9.

54 **Pennsylvania Avenue paved with Neuchatel rock asphalt:** Geo. C. Warren, "The Evolution of Asphalt Pavements in the United States," *Good Roads,* June, 1926.

54 **Pennsylvania Avenue—1832 macadamized, 1849 cobblestones, 1871 wooden blocks:** Cable, *The Avenue of the Presidents,* pp. 144, 146.

55 **"a surface as smooth as a billiard table . . .":** Dyer, Frank L., "The Streets and Roads of Washington," p. 318.

55 **"Washington is today universally known as the best paved . . . ":** "Asphalt Pavements," 1888, p. 9.

55 **H. L. Mencken, "The greatest of them . . . ":** *Happy Days 1880-1892,* John Hopkins University Press, 1996, p. 234.

55 **"Since the surface of these asphalt pavements . . . ":** Dyer, "The Streets and Roads of Washington," pp. 324, 327.

CHAPTER FIVE
THE ASPHALT TYCOON

56 **"Having given thirty years . . . ":** Amzi Lorenzo Barber, *The Best Road and the Right Way to Make It,* A. L. Barber Asphalt Company, New York, 1909.

56 **Oberlin College and board of Howard:** www.howard.edu/longwalk/ and link at www.howard.edu/welcomecenter.asp.

56 **land turned over to D.C. and became center of African American life:** www.exploredc.org/index.php?id=83

56–58 **"It was the inquiry conducted by the United States Government . . . "; and when company incorporated:** *The New York Times,* April 19, 1909, "Amzi L. Barber Dies of Pneumonia."

58 **challenged them with a ten-year warranty:** I. B. Holley, Jr., "Blacktop: How Asphalt Paving Came to the Urban United States," paper, The Society for the History of Technology, 2003, pp. 11, 12.

58 **delivered asphalt for less than half the cost of competitors:** Holley, "Blacktop: How Asphalt Paving Came to the Urban United States," pp. 10–11.

60 **negotiations for Indian reservation, two million versus 200,000 acres, presidential politics:** Holley, "Blacktop: How Asphalt Paving Came to the Urban United States," pp. 8–9.

61 **three-month-plus voyage from East Coast:** Raymond A. Rydell, *Cape Horn To The Pacific,* University of California Press, Los Angeles, 1952, p. 139.

61 **covered 80,000 miles of ocean and 400,000 miles over land during first four years:** Henry Hall ed., *America's Successful Men of Affairs: An Encyclopedia of Contemporary Biography,* New York Printing Co., New York, 1895, pp. 57–58.

61 **24 million yards of asphalt:** Holley, "Blacktop: How Asphalt Paving Came to the Urban United States," p. 10.

61–62 **$50 million worth of asphalt in 80 cities, 30 other companies laid half; net worth estimated at $7 million; "One of the greatest . . . ":** Hall, *America's Successful Men of Affairs,* pp. 57–58.

62 one of the first steam-powered yachts: *The New York Times*, April 19, 1909.

62 bought "factory, patent and parts": Floyd Clymer, *Treasury of Early American Automobiles*, McGraw-Hill Book Company, Inc., New York, 1950, p. 180.

63 law degree from Columbian University: Hall, *America's Successful Men of Affairs.*

63 Privy Court in London; "It seemed as though . . . ": *The New York Times*, April 19, 1909.

63–64 monopoly threatened; "inferior material": Holley, "Blacktop: How Asphalt Paving Came to the Urban United States," p. 9.

64 borrowed from Land Title and Trust Company: *The New York Times*, December 29, 1901.

64 controlling interests in nearly 80 percent of asphalt companies: Holley, "Blacktop: How Asphalt Paving Came to the Urban United States."

65 use by Aborigines; Largo La Brea—the Lake of Pitch; "impossible to estimate . . . ": *The New York Times*, February 17, 1901.

66 description of Guanaco: *The New York Times*, February 17, 1901.

66 "The asphalt . . . the finest and purest . . . "; "The refining process . . . ": *The New York Times*, February 17, 1901.

66 biscuit salesman; Hamilton and New Yorker struck a deal: *The New York Times*, July 26, 1908.

67 "a bandit in the office": *The New York Times*, December 20, 1908.

67 alleged $30 million embezzled: *The New York Times*, July 26, 1908.

67 "South America for South Americans": *The New York Times*, July 26, 1908.

67 military career of Major General Francis V. Greene: *The New York Times*, September 4, 1898.

67 revolutionaries financed by asphalt company; General Manuel Matos: *The New York Times*, October 19, 1905.

67 Green meeting Matos in Paris; shipyards in London, and Glasgow; "as the results of Greene's work in Europe . . . ": *The New York Times*, April 5, 1905.

68 Greene at State Department: editorial, *The New York Times*, January 8, 1901.

68 Barber forced out; 1903 new General Asphalt Company, absorbed 54 of 69 firms; "the operating arm": Holley, "Blacktop: How Asphalt Paving Came to the Urban United States," p. 11.

68 Congress, Venezuela, and Roosevelt involvement: *The New York Times*, May 23, 1905.

69 expense of war in Venezuela, default on loan: *The New York Times*, December 29, 1901.

69 Barber's death: *The New York Times*, April 19, 1909.

CHAPTER 6
"ARTIFICIAL" ASPHALT

71 "If we did not have asphalt . . . ": "Highway Builder's Yarn," *Asphalt Paving Technology*, Vol. 43-A, 1974, p. 76.

71 Barber Asphalt's survey: *Asphalt Paving Technology*, pp. 90–91.

72 driving at night; stopping by running at curb: Richard Weingroff, *A Partnership That Makes A Difference*, Washington, D.C., 2003, p. 78.

73 "Statistics in former years show . . . ": Amzi Lorenzo Barber, *The Best Road And The Right Way To Make It*, A.L. Barber Asphalt Company, New York, 1909: pp. 27–28.

74 Lamson & Bro. ad: T. Hugh Boorman, *Asphalts: Their Sources and Utilizations*, Asphalt for Dustless Roads, William T. Comstock, New York, 1908, p. 1.

75 Union Oil sent "around the Horn": author's phone interview with Don Nielsen, of Peoria, Arizona, September 7, 2004.

76 prices per yard of $3.36, $1.52: author's phone interview with Don Nielsen, of Peoria, Arizona, September 7, 2004.

77 artificial asphalt as petroleum: *Asphalt Paving Technology*, p. 70.

78 99.5 percent of asphalt from refineries unused: *Asphalt Paving Technology*, pp. 26–27.

78 girl oil dippers: *Asphalt Paving Technology*, p. 41.

80 Jesse Warren's trip to California; "making inroads into asphalt paving business in the East": George C. Warren, "The Evolution of Asphalt Pavements in the United States," *Good Roads*, June 1926, pp. 228–229.

80 "[T]here is nothing either to fear . . . ": George C. Warren, "The Evolution of Asphalt Pavements in the United States," pp. 228–229.

80–81 Alcatraz Asphalt Company; "dunes of oil"; pipeline effort a bust; real estate turned oil-exploration firm: George C. Warren, "The Evolution of Asphalt Pavements in the United States," pp. 228–229.

81 took barrels back to Boston: George C. Warren, "The Evolution of Asphalt Pavements in the United States," pp. 228–229.

81 "He discovered what no one else had dreamed . . . ": Herbert M. Warren, "The Warren Story," The Warren Brothers Company, 1968, quote from A.J. Allhands article in *American Builders*, 1954.

81 meeting recorded in family Bible: Herbert M. Warren, "The Warren Story," p. 2.

82 side hill plow; still in use in 1960s; fire and move to Glenn, NY: Herbert M. Warren, "The Warren Story," pp. 1, 2.

82–83 brothers have own businesses; offices around country; largest consumers of coal tar: Herbert M. Warren, "The Warren Story" p. 2.

83 coal-tar-naphtha "gained large profits": George C. Warren, "The Evolution of Asphalt Pavements in the United States," pp. 228–229.

83 Cyrus Warren's professorship "consumed too much of his time": Herbert M. Warren, "The Warren Story," p. 2.

84 John rebuild, sell to Standard Oil: Herbert M. Warren, "The Warren Story," p. 2.

85 Herbert Warren's death at sea; family retreat at his estate: Herbert M. Warren, "The Warren Story," p. 3:

86 patent of June 4, 1901; asphalt thickness: *Asphalt Paving Technology*, p. 175.

86 Bitulithic Macadam, scientifically proven consistency, first used in Pawtucket, R.I., in 1901: G. H. Perkins, "Bitulithic Pavement and Warrenite Roadway," *Boston Society of Engineers*, No. 3, March 1914, p. 119.

86 two-inch or three-inch layer: *Asphalt Paving Technology*, p. 175.

86 stones equal half depth of road; 60 percent more stone than asphalt; Perkins, "Bitulithic Pavement and Warrenite Roadway," p. 121.

86–87 "It is impossible for us to formulate a standard . . . ": Perkins, G. H., "Bitulithic Pavement and Warrenite Roadway," p. 121.

87 Fred Warren died at 39 in hotel room; patent lasted until 1918: Herbert M. Warren, "The Warren Story," pp. 4, 5.

89 first patent July 31, 1790; signatures of Washington and Jefferson: George Warren, lecture delivered before the graduate students in highway engineering at Columbia University on February 12, 1915, Columbia University, 1914, quoting Charles Allen Munn, managing editor of *Scientific American,* in his article "The Laws of Patents."

89 "I have invented certain new and useful . . . ": Hugh M. Gillespie, *A Century of Progress,* National Asphalt Pavement Association, Maryland, 1992, p. 75.

89 quote from *Kendall* v. *Winsor* "entitled to protection . . . ": George Warren, lecture, February 12, 1915.

90 Warren patent upheld; "happy to have it be general knowledge . . . ": Richard L. Davis, *Asphalt Pavements,* Richard L. Davis, Pittsburgh, 2003, p. 24.

90 first Bitulithic plant built next to Cambridge Gas Works: Herbert M. Warren, "The Warren Story," p. 4.

90 1907 imports of natural asphalt dropped by 80 percent: I. B. Holley, Jr., "Blacktop: How Asphalt Paving Came to the Urban United States," paper: The Society for the History of Technology, 2003, p. 22.

91 the chew test: *Asphalt Paving Technology,* p. 93.

92–93 suit against Kaw Paving; the Topeka Decree May 26, 1910; stones smaller than one half inch; 90 percent pass through openings of one quarter inch: Gillespie, *A Century of Progress,* p. 79.

93 "raisins in a fruit cake": Herbert M. Warren, "The Warren Story," p. 4.

93 Amiesite, Willite, Romanite, etc: www.hotmix.org/view_article.php?ID=69

94 percentages of paved surfaces per mid-1920s survey; asphalt better for horses; easier to repair: Holley, "Blacktop: How Asphalt Paving Came to the Urban United States," p. 30.

94–95 asphalt's success due to Barber and Warren companies: Holley, "Blacktop: How Asphalt Paving Came to the Urban United States," p. 30.

CHAPTER 7
HOT MIXING, RAKING, AND ROLLING

97 "By and large, the difference . . . ": James Morris, quoted by Christy Borth, *Mankind on the Move,* Automotive Safety Foundation, Washington, D.C., 1969, p. 165.

98 1849, M. Merian: *Asphalt Paving Technology, Vol. 43-A,* 1974, p. 160.

98 lay-down procedure; waiting before compacting with rollers: *Asphalt Paving Technology,* pp. 162, 164.

99 1863 Aveling Porter steamroller; not used on rock asphalt: *Asphalt Paving Technology,* pp. 161, 162.

99 1871 patent for asphalt to N. B. Abbot; "I have . . . ": *Asphalt Paving Technology,* p. 163.

101 Abbott's equipment developments: *Asphalt Paving Technology,* p. 163.

101 four hours to mix a batch: Hugh M. Gillespie, *A Century of Progress,* National Asphalt Pavement Association, Maryland, 1992, p. 81.

101 rotating drum within stationary one; convex bottom box: *Asphalt Paving Technology,* pp. 163–164.

101 Pioneer Iron-Works: *Asphalt Paving Technology,* p. 164.

101-102 "That Abbott did not understand proportioning . . . ": *Asphalt Paving Technology,* p. 164.

102 1862 William Barford's horse-drawn rollers: Gillespie, *A Century of Progress,* p. 42.

103 Aveling and Porter steamroller; description of hand roller: *Asphalt Paving Technology,* p. 174

104 $10 versus $20 a day for horses: Gillespie, *A Century of Progress,* p. 64.

104 drums too narrow; 1869 John Roach; 1875 Andrew Lindelhof's 10-ton roller: *Asphalt Paving Technology,* p. 164.

104 Americans' distinctive use of asphalt rollers: *Asphalt Paving Technology,* p. 164.

105 1908 Barford and Perkins first solid-injection diesel engine: *Asphalt Paving Technology,* p. 190.

105 Buffalo Springfield three-axle rollers: Gillespie, *A Century of Progress,* p. 64.

107 Barber not believed to make own plants: *Asphalt Paving Technology,* p. 170.

106–107 Barber's and Europe's production levels: *Asphalt Paving Technology,* p. 202.

107 two Warren patents on April 14, 1903: *Asphalt Paving Technologies,* p. 174.

108 10,000 gallons, nearly pure: "Socony Handbook," Standard Oil Company of New York, 1932, p. 5.

110 1899, 60-foot railcars: *Asphalt Paving Technology,* p. 188.

110 By 1920, 80 semiportable plants: *Asphalt Paving Technology,* p. 201.

110 July 18, 1913 patent: *Asphalt Paving Technology,* p. 194.

111 Benjamin Holt tested in 1904: William R. Haycraft, *Yellow Steel: The Story of Earthmoving Equipment Industry,* University of Illinois, Chicago, 2000, p. 49.

111–112 patent July 8, 1913; Warrenite portable plants described: *Asphalt Paving Technology,* p. 195.

112 one cubic yard; three tons an hour: *Asphalt Paving Technology,* p. 194.

113 dual-chamber mixer; pave roads in remote areas: Gillespie, *A Century of Progress,* p. 93.

113 gob boxes: *Asphalt Paving Technology,* p. 201.

114 small loads of asphalt: *Asphalt Paving Technology,* p. 174.

114 In 1899 . . . 300 tons to New Rochelle; hot mixes could be stored: *Asphalt Paving Technology,* pp. 169, 203.

114 advantage of trucks: *Asphalt Paving Technology,* p. 203.

114 dump board; wagon dumper: *Asphalt Paving Technology,* p. 206.

CHAPTER EIGHT
ATTACKING ASPHALT WITH TANK AND TRUCK

118 "Unclassable . . . ": Good Roads activists jingle, from Paul Dickson and William Hickman Forbes, *Firestone Centennial,* Bridgestone/Firestone Tires Inc., New York, 2000, p. 71.

119 since 13th century, trucking has meant . . . : *Merriam-Webster's Collegiate Dictionary,* tenth edition, 2000.

121 San Francisco earthquake . . . confiscation of 200 private trucks and automobiles; 15,000 gallons of fuel donated: James J. Flink, *America Adopts the Automobile,* MIT Press, Cambridge, Massachusetts, 1970, p. 45.

121 Pioneer Freighter: *America's Highways 1776–1976,* U.S. Department of Transportation Federal Highway Administration, Washington, D.C., 1976, p. 62

122 modern warhorses—motortrucks: Robert F. Karolevitza, *This Was Trucking*, Bonanza Books, New York, MCMLXVI, p. 61.

122 "When your trucks get stuck . . .": Karolevitza, *This Was Trucking*, p.52.

122 "Road so bad . . . ": Karolevitza, *This Was Trucking*, p. 53.

123 France used 8000 trucks: Dickson and Forbes, *Firestone Centennial*, p. 73.

124 "Ship by Truck": Dickson and Forbes, *Firestone Centennial*, pp. 70–71.

125–126 September of 1918; Wingfoot Express; cotton-cord fabric; pneumatic tires: Jefferey Rodengen, *The Legend of Goodyear*, Write Stuff Syndicate, Fort Lauderdale, Florida, 1997, p. 48.

126 army ordered 30,000 trucks: *America's Highways 1776–1976*, p. 95.

126 Pershing needed 50,000 trucks and endless parts: Karolevitza, *This Was Trucking*, p. 62.

127 trucks with Detroit to Berlin banners; deep snow: pictures in *America's Highways 1776–1976*, pp. 95, 96.

128 freed 17,250 boxcars; ammunition and spare parts trucked: *America's Highways 1776–1976*, p. 95.

129 "In those days we were not sure . . . ": Dwight D. Eisenhower, *At Ease: Stories I Tell To Friends*, Doubleday & Company, 1968, p. 157.

130 "This is the beginning of a new era"; "Proceed by way of the Lincoln Highway . . .": Richard Weingroff, "The Man Who Changed America," *Public Roads*, FHWA, March/April 2003, p. 22.

130 "existed largely in the imagination and on paper": Captain William C., U.S. Army Memorandum, July to September 6, 1919.

132 "Those who do not have . . . ": Tom Lewis, *Divided Highways*, Viking Penguin, New York, 1997, p. 13.

132 more work than in the previous three years combined; approved 90 percent of state highway requests: Bruce Seely, *Building the American Highway System*, Temple University Press, 1987, p. 55.

132 The Chief didn't stop there. . . . in the construction of his new interstate highways: *America's Highways 1776–1976*, pp. 105-106.

133 Pershing and his map: *America's Highways 1776–1976*, p. 142.

133 Napoleon; despots and democracy: Lewis, *Divided Highways*, p. 20.

CHAPTER 9
LYING LIGHTLY ON THE LAND

135 "We pioneered a new road . . . ": phone interview between author and George Wagner, August 11, 2005.

135 "The public would follow any road . . . ": www.nbm.org/blueprints/summer97/page2/page2.htm

135 "lie lightly on the land": www.cr.nps.gov/habshaer/lll/post/

135 1924 to 1933 building Going-to-the-Sun Road: www.cr.nps.gov/habshaer/lll/post/glacier.htm

135–136 52-mile road; mountainsides, glacial lakes and arctic tundra; Continental Divide at 6646 feet; viewpoints: www.visitmt.com/

136 access before Going-to-the-Sun was built: www.glacier.national-park.com/info.htm

136 Golden Age of park roads: www.cr.nps.gov/habshaer/lll/post/

136 "At the beginning of the survey . . . ": ww.cr.nps.gov/nr/twhp/wwwlps/lessons/95sunroad/95facts1.htm

136-137 "With masonry guardrails . . . ": montanakids.com/

138 model for road construction in all the national parks: www.cr.nps.gov/habshaer/lll/post/

138 "one of the crown jewels . . . ": www.house.gov/apps/list/press/mt00_rehberg/011304_HighwayGTSR.html

138 National Register of Historic Places and National Historic Civil Engineering Landmark: www.visitmt.com/

138 National Historic Landmark in 1997: e-mail from Timothy Davis, Ph.D., Lead Historian, Park Historic Structures & Cultural Landscapes Program, U.S. National Park Service, to Margaret Cervarich, September 19, 2005.

138 June 23, 1900, Lippincott first; Locomobile; reduced stagecoach trip; hundreds of thousands of autos: Ellie Huggins, *All Roads Lead to Yosemite*, Cold Stream Press, Truckee, California, 2002, p. 22.

138 John Muir: Huggins, *All Roads Lead to Yosemite*, p. 20.

139 1926 all-year highway: Huggins, *All Roads Lead to Yosemite*, p. 22.

139–142 reminiscences of George Wagner: Granite Construction Company's 75th Anniversary publications and phone interview between author and Wagner, August 11, 2005.

142–143 rock for building the road was quarried in the park; now use local aggregate of appropriate color: phone call between Margaret Cervarich and Timothy Davis, National Park Service historian, September 15, 2005.

CHAPTER TEN
ASPHALTING THE MOTHER ROAD: ROUTE 66

144: "The highway to San Juan . . .": John Steinbeck, *The Wayward Bus*, The Viking Press, New York, 1939. p. 160.

146 California concrete highways had to be replaced: Susan Croce Kelly, *Route 66: The Highway and Its People*, University of Oklahoma Press, 1988, p. 31.

148 "The miracle was not the automobile. . . .": Kelly, *Route 66: The Highway and Its People*, p. 3.

148 "Not many people traveled . . ." and other quotes from Michael Burns: Kelly, *Route 66: The Highway and Its People*, p. 18.

148 "It was not Highway 66 . . ." and other quotes from George Greider: Kelly, *Route 66: The Highway and Its People*, p. 20.

149 "If two automobiles met . . .": Kelly, *Route 66: The Highway and Its People*, p. 20.

149 "My aim is this . . .": Tom Lewis, *Divided Highways*, Viking Penguin, New York, 1997, p. 19.

151 patents began to expire; federal and state engineers making findings public: Hugh M. Gillespie, *A Century of Progress*, National Asphalt Pavement Association, Maryland, 1992, p. 43.

151–153 James Francis Gallagher's career: Asphalt Association, Circular No. 9, New York, no date (mid 1920s); author's phone interview with James's son, Jack Gallagher, in Flossmoor, Illinois, May 3, 2005; Margaret Cervarich's phone interview with James's grandson, Dan Gallagher, May 26, 2005.

153 "Route 66 is a giant chute . . .": James R. Powell, former president of Missouri Route 66 Association, unpublished paper, "A Brief History of Route 66," May 6, 2002, p. 1.

154 Avery as real estate speculator, entrepreneur, car dealership and tourist court: Kelly, *Route 66: The Highway and Its People*, p. 5.

156 "[I]f the development of the highway continues . . .": Kelly, *Route 66: The Highway and Its People*, p. 24.

156-157 "The first time I went into Needles . . .": Kelly, *Route 66: The Highway and Its People*, p. 20.

157 "In those days Highway 66 . . .": Kelly, *Route 66: The Highway and Its People*, p. 57.

157 U.S. 66 Association's ad; "You stay on pavement longest going west . . ."; 700 responded: Kelly, *Route 66: The Highway and Its People*, p. 36.

157 Los Angeles population doubled: Kelly, *Route 66: The Highway and Its People*, p. 31.

158 in 1927, Griffith company first with automated paver: *Asphalt Paving Technology*, Vol. 43-A, p. 218.

158 Cannady tries concrete finisher: *Asphalt Paving Technology*, p. 219.

159 50 percent increase in the speed of laydown: *Asphalt Paving Technology*, p. 218.

159 mechanical pavers required in California: *Asphalt Paving Technology*, p. 221.

159 in 1928, 820 tons in a day: *Asphalt Paving Technology*, p. 222.

159 "With virtually only two years . . .": Gillespie, quoting *Engineering News Record* in *A Century of Progress*, p. 50.

160 "If it weren't for the floating screed . . .": NAPA Midyear Meeting, Chicago, Illinois, August 2004.

160 the 1930 Road Show did not have steam shovels: Charles W. Wixom, *A Pictorial History of Road Building*, American Road Builders Association, Washington, D.C., p. 102

160 "the first functional bituminous concrete paver"; Barber's background: *America's Top 100 Private Sector Transportation Design & Construction Professionals of the 20th Century*, American Road & Transportation Builders Association Program, October 18, 2004, Washington, D.C., p. 9.

160 Greene's background: American Road & Transportation Builders Association Program, October 18, 2004, Washington, D.C., p. 39.

160 Sheldon G. Hayes first to use floating screed, in Rolla, Missouri: Gillespie, *A Century of Progress*, p. 50.

162 dust storm stripped 300 million tons of topsoil; "Those years of 1934 and 1936 . . .": Kelly, *Route 66: The Highway and Its People*, p. 57.

162 "Dad, he built roads. . . .": Kelly, *Route 66: The Highway and Its People*, p. 28.

162 "starved to death. . . "; "We moved to town . . .": Kelly, *Route 66: The Highway and Its People*, p. 29.

163 "The Grapes of Wrath went through here . . .": Kelly, *Route 66: The Highway and Its People*, pp. 59–60.

164 "When they paved this road . . .": Kelly, *Route 66: The Highway and Its People*, p. 67.

164 "When the Okies got here . . .": Kelly, *Route 66: The Highway and Its People*, p. 62.

164 August 1938, final paving of Route 66 between Adrian and Glenrio: Jim Powell, *Show Me Route 66*, Vol. 12, No. 4, Fall 2001, p. 11.

165 Thew made first crane; "strength of 12 men"; Hi-way Yellow: William R. Haycraft, *Yellow Steel: The Story of Earthmoving Equipment Industry*, University of Illinois, Chicago, 2000, p. 64.

166 trucks with 1000-gallon tanks; two and one half gallons of bitumen per square yard; avoid leaving "fat and lean" spots: *Asphalt Paving Technology*, p. 240.

166 "Tarvia, Better Roads at Lower Costs" became the slogan . . .: "The First Hundred Years," unpublished paper of The Barrett Company, 1954, p. 7.

166 a household name and in Webster's Dictionary: Donald Rosenbarger, P.E., MWRA Interchange, newsletter of the Michigan Road Builders Association, Winter 2004, p. 19.

166 in 1926, first mix-in-place section laid in California: *Asphalt Paving Technology*, p. 213.

167 Butler tailgate spreader, 1924, used especially in southwestern United States: *Asphalt Paving Technology*, p. 217.

167 six major firms; Warren Brothers largest plant in the world: *Asphalt Paving Technology*, p. 239.

168 throwing flames and heat on the road. . . smooth patch of asphalt: *Asphalt Paving Technology*, p. 249.

167–169 "They just tore up 66 . . .": Kelly, *Route 66: The Highway and Its People*, p. 66.

169 "The only thing that brought us out of the Depression was . . .": Kelly, *Route 66: The Highway and Its People*, p. 66.

170-171 "Starting out, times were . . . " and further quotes about Tiffany Construction: author interview with Herb Tiffany, Jr., Phoenix, Arizona, December 29, 2004.

171 eight million Americans to the west, nearly half settled in California: Kelly, *Route 66: The Highway and Its People*, p. 148.

CHAPTER ELEVEN
PAVING THE WAY TO VICTORY

172 "Undoubtedly, the ability to build pavements quickly . . .": *Asphalt Paving Technology*, Vol. 43-A, 1974, p. 254.

172 $15.6 billion (over $180 billion in 2005 terms): Inflation conversion factor of .084 from oregonstate.edu/dept/pol_sci/fac/sahr/sahr.htm.

173 paving of Pennsylvania Avenue in Washington, D.C., with Trinidad lake asphalt: "The History of Asphalt—Part One," *Paving the Way*, Plant Mix Asphalt Industry of Kentucky and Kentucky Asphalt Paving Alliance, Frankfort, Kentucky, July 2002.

176 a "fifth rate air force" per General Andrews: United States Strategic Bombing Survey (Pacific War), Washington, D.C., 1946.

176 as late as 1939, sod surfaces would be adequate for advance bases: Blanche Coll, Jean Keith, and Herbert Rosenthal, "The Revolution in Equipment," *The Corps of Engineers: Troops and Equipment*, Washington, D.C., U.S. Army Corps of Engineers, 1975, pp. 56, 57.

177 manual for "Design of Airport Runways": Lenore Fine, and Jesse Remington, *The Corps of Engineers: Construction in the United States*, Washington, D.C., U.S. Army Corps of Engineers, 1972, p. 447.

177 a runway could be built during the night and used the next morning . . . : Letter from Richard G. Ahlvin to Alfred Chock, Jr., March 2004.

182 commission for Porter quashed: Fine and Remington, *Construction in the United States*, p. 636.

182 Hubbard-Field test . . . Hveem test: Campbell Crawford, "The Rocky Road of Mix Design," *Hot Mix Asphalt Technology* magazine, Winter 1989.

186 Seabees had no protections under the Geneva Convention: Vincent A. Transano, "Seabee History: Formation of the Seabees and World War II". Washington, DC: Department of the Navy, Naval Historical Center, 1997, at www.history.navy.mil/faqs/faq67-3.htm.

186 Atlantic and Pacific "roads to victory": Transano, "Seabee History: Formation of the Seabees and World War II".

186 route plan for attack on Japan: U.S. Navy, *Building the Navy's Bases in World War II: History of the Bureau of Yards and Docks and the Civil Engineer Corps, 1940–1946*, Volume II, 1947, Washington, D.C., Government Printing Office.

187 obstacle of freezing runway fill: *Building the Navy's Bases in World War II*, pp. 57–58.

189 over 100 planes a day could pass through Patterson and Meeks Fields: Mary Kegley, "Charles O. Fortson's Diary: A Glimpse into the Life of a 1939 Alumnus During World War II", *Citadel Pass in Review*, Nov./Dec. 2004, www.citadel.edu/pao/e_let/sy04-05/200411/a_diary.shtml

190 Seabees won 33 Silver Stars, five Navy Crosses; 300 died in combat and 500 in construction accidents : Transano, "Seabee History: Formation of the Seabees and World War II."

190 Seabees, etc., built 111 airfields in Pacific: Transano, "Seabee History: Formation of the Seabees and World War II."

190 use of "native materials"; permanent runways often asphalt: *Building the Navy's Bases in World War II*, pp. 370–373.

192 Saipan key to invading Japan: Air Force Historical Foundation, *IMPACT: The Army Air Forces' Confidential Picture History of World War II*, Published in 8 Books, Vol. 6, Historical Times, Inc., Harrisburg, Penna., 1982, pp. 22–31.

192 the tide of the war had turned: Philip A. Crowl, *U.S. Army in World War II: The War in the Pacific: Campaign in the Marianas*, Washington, D.C., Center of Military History, United States Army, 1993, p. 445.

193 capture of Iwo Jima and Seabees build airstrips: Wesley Craven, and James Cate, eds., *The Army Air Forces in World War II. Vol. V, The Pacific: Matterhorn to Nagasaki*, University of Chicago Press, Chicago, 1953, pp. 593–596.

196 story of the *Enola Gay*: Children of the Manhattan Project, "Timeline #2: The 509th: The Hiroshima Mission," www.childrenofthemanhattanproject.org/CG/CG_09C1.htm. Also www.456fis.org/B29.htm. And "Tinian in WWII - Teamwork and Effort," www.112thseabees.com/history/tin_hist.htm.

196 more than 1400 airfields built for the military: www.armyairforces.com/

196 500 airfields surplus or shut down: Scott Murdock, "The Use In 1995 of World War II Army Air Fields in The United States," graduate paper for Embry-Riddle Aeronautical University, published 1997, and on the Web at www.airforcebase.net/aaf/grphtml.html, updated August 2002.

198 "Congressman Hebert . . .": quoted in National Bituminous Concrete Association's *NBCA Newsletter*, vol. 2, no. 3, Washington, D.C., July 15, 1957.

198 weighed in at 300,000 pounds: www.globalsecurity.org. Note: estimates on the maximum gross take-off weight vary widely by variant and source.

198 maximum gross weight of over 200,000 pounds for B-47: www.globalsecurity.org. "Weapons of Mass Destruction-B-47 Stratojet."

198 President Truman was being "hounded" by asphalt and concrete proponents: Interview with General Robert Landry, Executive Officer to General Spaatz, from the Truman Library Web Site.

198 LeMay's policy for runways and aprons: *Engineer Memoirs: Major General William E. Potter, Retired, USACE*, Pub. 870-1-12, 1983.

199 "admitted to a lack of confirming evidence": Richard G. Ahlvin, "Origins of Development for Structural Design of Pavements," USACE, Waterways Experimental Station, Vicksburg, MS, Technical Report GL-91-26, pp. 4-15, 4-16.

199 December 1955, revised criteria prohibited the construction of asphalt runways: Ahlvin, "Origins of Development for Structural Design of Pavements," pp. 4-16.

199 subcommittee issued its findings and recommendations in March 1959, and ban lifted: Ahlvin, "Origins of Development for Structural Design of Pavements," pp. 2-12-2.13.

CHAPTER TWELVE
THE FIRST ASPHALT SUPERHIGHWAYS

202 "After careful consideration . . .": Paul L. Troast, chairman New Jersey Turnpike Authority, 1951, quoted by Herbert Spencer, "The New Jersey Turnpike: One of America's Heavy-Duty Highways," *Asphalt Quarterly*, October 1951, p. 7.

202 "After the war there was a trafic jam"; "One of the things . . .": Susan Croce Kelly, *Route 66: The Highway and Its People*, University of Oklahoma Press, 1988, p. 148.

202 Road-paving projects had been put on the back burner: *Asphalt Paving Technology*, Vol. 43-A, 1974, p. 252.

204 a time of "application and abandonment": *Asphalt Paving Technology*, p. 251.

204 "five winters and four years": *Asphalt Paving Technology*, p. 252.

204 "It was thin gruel . . .": author interview with Chan Rogers, March 2003.

204 1800 command installations and 2200 industrial facilities: *Asphalt Paving Technology*, p. 254.

205 jumped to 200 travel plants a year and 500 finishers; peacetime army of "engineers, technicians, supervisors, operators, mechanics,": *Asphalt Paving Technology*, p. 253.

205–206 "It was used in the South Pacific . . . the plant and the flag are still around" all quotes from Kemmetern: author interview with Peter Kemmetern of Heckert Construction in his office in Pittsburg, Kansas.

209 "the lanes were asphalt concrete but the shoulders . . .": e-mail of Roland Lavallee to Bruce D. Pelletier, forwarded to author from Maine Turnpike Authority, June 23, 2005.

210 its 240 bridges: New Jersey Turnpike, 1949 Annual Report.

210–211 "The turnpike has the distinction . . .": the city's "western doorstep": Armand Schwab, Jr., "City Linked To Superhighway," *The New York Times*, Sunday, January 20, 1952.

211 "The tortuous trip . . .": "America's Finest Highway: The New Jersey Turnpike," *Road International*, Spring, 1952.

211 the Corridor State: Spencer, *Asphalt Quarterly*, October 1951, p. 6.

211 "Geographically New Jersey is . . .": John Dietz, *Civil Engineering*, January 1952, p. 26.

211 "America's finest highway . . . even surpassing . . .": *Road International*, Spring, 1952.

212 "The first meeting of the engineering firms . . .": Dietz, *Civil Engineering*, January 1952, p. 72.

213 "Asphalt Penetration Macadam Base": *Road International*, Spring, 1952.

213 concrete pavement . . . $45,729,568 . . . an asphalt highway $40,272,434: *Road International,* Spring, 1952.

214 $12,000-only $64 per year per mile: Dietz, *Civil Engineering,* January 1952, p. 32.

214 concrete-slab replacements; "The wartime history . . .": Spencer, *Asphalt Quarterly,* October 1951, pp. 6-7.

215 "One of the most eminent . . . "; "After careful consideration . . .": Spencer, *Asphalt Quarterly,* October 1951, p. 7.

215 700 engineers . . . 10,000 construction workers . . . $45 million: *Road International,* Spring 1952.

215 veterans of the "war period": Spencer, *Asphalt Quarterly,* October 1951.

215–216 supplying 17 Barber-Greene and two Adnun pavers; 175 tons of asphalt; 10-ton rollers: Leslie M. Stewart, *Civil Engineering,* 1952, p. 79.

216–217 14 million gallons of asphalt binder; 11.3 million gallons of asphalt; daily 15,000 tons of aggregate, 160,000 of asphalt: Spencer, *Asphalt Quarterly,* October 1951.

219 range 5.25 to 5.8 pounds; twice the amount of weight on most turnpikes: Spencer, *Asphalt Quarterly,* October 1951.

219–220 side form requirement dropped; change from waterbound macadam to asphalt penetration macadam; 3.2 gallons per square yard; haul roads; work on drains: Dietz, *Civil Engineering,* January 1952, p. 76.

221 the 6.3-million-square-yard job; 4000 . . . 5000 . . . 6000 tons per day; 120 tons per hour: Spencer, *Asphalt Quarterly,* October 1951.

226 "the method of feed . . .": Dietz, *Civil Engineering,* January 1952, p. 78.

226 one-eighth-inch variance: Spencer, *Asphalt Quarterly,* October 1951, p. 13.

226 with the top down, the radio on . . . : reminiscence of the author's father, Dan McNichol, about the newly constructed New Jersey Turnpike.

229 In the 10 years after . . . 45 different . . . : Bureau of Public Roads, "Report on Arterial Toll Roads, Toll Parkways and Resort Toll Roads," January 1, 1955

229 "its joining with the Pennsylvania Turnpike . . .": "Ohio's 241-Mile Turnpike," *The New York Times,* September 25, 1955, pp. 21–22.

230 "the road that couldn't be built"; "Road Nowhere"; "On a map . . ."; "The first step . . .": "Road Nowhere, A Link South," special to *The New York Times,* November 14, 1954, p. 54.

230 October 21, 1956; Gene Autry and Champion; Wyoming governor: Sherry Lamb Schirmer, and Dr. Theodore A. Wilson, *Milestones: A History of the Kansas Highway Commission.*

231 a year and a half later . . . I-35: William Paul Corbett, "Okalahoma Highways: Indian Trails to Urban Expressways."

CHAPTER THIRTEEN
IKE'S GRAND PAVING PLAN

232 "Americans are living in the midst of a miracle. . . .": Robert Paul Jordan, "Our Growing Interstate Highway System," *National Geographic,* February 1968, pp. 194–195.

232 longest standing ovation: Stephen Ambrose, *Eisenhower: Soldier and President,* Simon & Schuster, 1990, p. 207.

234 In 1949 . . . nearly 40,000 miles of highway had been audited by the PRA . . . ; average age 12 years: 1949 PRA report to Congress, "Highway Needs of the National Defense."

235 "comparable to the casualties . . . "; "appalling inadequacies . . . "; "I would like to read to you the last sentence . . .": Richard Weingroff, "The Man Who Changed America: Part 1" *Public Roads,* March/April 2003, FHWA, Washington, D.C., p. 29.

236 "A modern, efficient highway system . . .": Weingroff, "The Man Who Changed America: Part 1," p. 33.

237 estimated 70 million people; army . . . requested beltways: Tom Lewis, *Divided Highways,* Viking Penguin, New York, 1997, p. 108.

237 "Each year more than 36,000 . . .": White House press release, February 22, 1955.

237 "The country urgently needs . . .": Weingroff, "The Man Who Changed America, Part II," *Public Roads,* May/June 2003, p. 23.

238 Hayes purchased the first Barber-Greene floating screed: *America's Top 100 Professionals of the 20th Century,* American Road & Transportation Builders Association, Washington, D.C., 2004, p. 44.

238–239 about Sheldon G. Hayes; "was down at the speedway . . .": Sheldon W. Hayes, phone interview by Margaret Cervarich, July 11, 2004.

240 Portland Cement Association and the American Concrete Pavement Association pouring millions into a campaign: Margaret Blain Cervarich, *Fifty Years of NAPA,* 2005, NAPA, Washington, D.C., p. 3.

240 "The boys feel that a National Association. . . .": Minutes of the Board of Directors meeting, National Bituminous Concrete Association, 4/26–27/55, Conrad Hilton Hotel, Chicago.

240 lecture given by Charles M. Noble: National Bituminous Concrete Association's *NBCA Newsletter,* vol. 1, no. 1, Washington, D.C., March, 1956.

241 over 1.6 million acres of land: Jordan, "Our Growing Interstate Highway System," p. 198.

243 threefold increase in asphalt 1950-1960, other increases in next two decades: *Asphalt Paving Technology,* 1974, pp. 261 and 273.

243 C. W. "Duke" Beagle began experimenting with thick bases: *Asphalt Paving Technology,* p. 274.

244 "A parachute flare . . .": American Road Builders Association's *American Road Builder* magazine, November 1958, p. 10.

245 using dump trucks with double-axles in rear: *Asphalt Paving Technology,* p. 260.

245 attached to the back of semitrailers: *Asphalt Paving Technology,* p. 268.

249 in 1973, Barber-Greene manufactured a "super plant": *Asphalt Paving Technology,* picture on p. 286.

248–249 "Back in the early '60s . . .": Margaret Cervarich phone interview with Tom Ritchie, August 2005.

249–250 "Our customers knew . . . " etc.: author phone interview with Dr. J. Don Brock, in Tennessee, July 16, 2005.

250 recording pyrometers used in 1935, later required by California: *Asphalt Paving Technology,* p. 236.

250 by the 1970s, electronics in asphalt plants: *Asphalt Paving Technology,* p. 276.

250 bills generated electronically: *Asphalt Paving Technology,* p. 287.

251 "In the West Indies . . .": e-mail from Roger Sandberg to Margaret Cervarich, September 9, 2005

251–252 "Competition is what makes America great. . . ." etc.: author's phone interview with Chuck Deahl in Illinois, May 27, 2005.

252 "Rubber tires allowed gear . . ." etc.: author's interview with Bart Mitchell at NAPA's 50th Anniversary Convention, Hawaii, February 15, 2005.

253 "In 1963, a 16-foot-wide asphalt paver cost $14,000. . . .": author's interview with Bart Mitchell, February 15, 2005.

254 "Pneumatic-tired rollers . . ." etc.: author's phone interview with Tom Skinner in Illinois, May 27, 2005.

255 "Almost 30,600 miles . . .": press release: USDOT, Federal Highway Administration, December 7, 1970.

256 "the traffic that does come . . .": *NAPA Newsletter,* April 1962, p. 2

257 "Perhaps as a by-product . . .": William R. Haycraft, *Yellow Steel,* University of Illinois Press, Chicago, 2002, p. 187.

258 gasoline from 30 cents a gallon to as much as $1.20; asphalt and Dow Jones declined: Cervarich, *Fifty Years of NAPA,* p. 11.

258 a cost of roughly $27 billion estimated; less than 500 miles a year; final total more than $130 billion: USDOT press releases, according to a 1993 USDOT-FHWA brochure.

259 "As the Supreme Commander of the Allied Forces, Eisenhower achieved . . .": author's interview with former Massachusetts State Senator John M. Quinlan, June 30, 2005.

CHAPTER FOURTEEN
PAVING THE WAY TO THE FUTURE

260 "Hot mix asphalt technology . . .": John Becsey, Michigan Asphalt Paving Association, quoted in the Asphalt Pavement Alliance's *Alliance in Action* newsletter, November 19, 2001.

262 "I made a literature search . . .": letter on NAPA letterhead from Charles Foster to James Gallagher, October 15, 1965.

262 "Liquid asphalt in 1972 . . .": Margaret Cervarich's phone interview with Bill Swisher, August 26, 2005.

263 "The most expensive part . . .": interview with Swisher, August 26, 2005.

264 "I took the idea home . . .": videotaped interview with Dick Sprinkel, Newport Beach, California, November 8, 2004.

266 K. E. McConnaughey pioneered a method: Gillespie, *A Century of Progress,* National Asphalt Pavement Association, Maryland, 1992, p. 95.

267–268 The $10-Million Question: Prithvi (Ken) Kandhal, *The History of the National Center for Asphalt Technology,* National Center for Asphalt Technology, Auburn, Alabama, 2004.

270 word reached the U.S. from Germany: Margaret Cervarich's interview with Mike Acott, Lanham, Maryland, July 5, 2005.

270–273 "The Right Thing To Do": Margaret Cervarich's phone interviews with Gail Mize, Joe Musil, Ted Rapallo, and Paul Schulz; and interview with Gary Fore, Lanham, Maryland; all, February 22–27, 2002.

273–277 Testing the Tests: E. Ray Brown, Ph.D., "Scoring at Home," *Roads & Bridges* magazine, January 2003.

278 "Rubblization makes the base into. . . .": Bob Walters, quoted by Farrell Wilson, "Arkansas' Interstate Rehabilitation Program," *TR News,* March–April 2002.

280–281 "Believers Along Superstition Freeway": Mike Fickes, "The Asphalt Rubber Phenomenon," *Hot Mix Asphalt Technology* magazine, July/August 2003, pp. 20–23.

282 "If we can make this technology work in the U.S. . . .": videotaped interview with David Carlson, Washington, D.C., September 24, 2002.

282 "It is difficult for many . . .": Prepared remarks of Charles F. Potts, NAPA's 49th Annual Convention, Phoenix, Arizona, September 19, 2004.

BIBLIOGRAPHY References consulted by the author of Chapter Eleven, "Paving the Way to Victory"

Ahlvin, Richard G. *Origins of Developments for Structural Design of Pavements.* Vicksburg, MS: USACE Waterways Experiment Station, Geotechnical Laboratory, 1991.

Ahlvin, Richard G. Research interview. March 22, 2004.

Ahlvin, Richard G. "The Response to Developing Aircraft Load Support Needs" (handwritten notes). March 2004.

Baugher, Joe. "Bomber Series-Douglas B-19."
From www.ibiblio.org/pub/academic/history/marshall/military/airforce/.

Bituminous Pavements Standard Practice. Chapter 9 in Army Technical Manual TM-522-8, Air Force Technical Manual AFM 88-6, 1987, 1992.

Chandler, Ira. "1823-First American Macadam Road."
From http://curbstone.com/_macadam.htm.

Coakley, Robert W. *American Military History.* Chapter 23 in "World War II: The War Against Japan." Washington, D.C.: Center of Military History, 1989.

Collins, John M. "Lines of Communication." In *Military Geography for Professionals and the Public.* Washington, D.C.: Institute for National Strategic Studies, National Defense University, Fort McNair, March 1998.

Davies, Pete. *American Road—The Story of an Epic Transcontinental Journey at the Dawn of the Motor Age.* New York: Henry Holt & Company, 2002.

Dunn, Carroll H., Lt. Gen. *Vietnam Studies: Base Development in Vietnam, 1965–1970.* Fort Belvoir, VA: Office of History, U.S. Army Corps of Engineers, 1972.

Hecker, Siggerud, et al. "Highway Infrastructure: Interstate Physical Conditions Have Improved, but Congestion and Other Pressures Continue." GAO Report 02-GAO-571. Washington, D.C.: General Accounting Office, June 2002.

Hendricks, Charles. "The Air Corps Construction Mission." In Barry W. Fowle, ed., *Builders and Fighters: U.S. Army Engineers in WWII.* Pub. EP 870-1-42. Fort Belvoir, VA: Office of History, U.S. Army Corps of Engineers, 1992.

"History of 509th Composite Group 313th Bombardment Wing, Twentieth Air Force, Activation to 15 August, 1945—Tinian in the Marianas." Maxwell AFB, Georgia: Aerospace Studies Institute, Archives Branch, 1945.

McDonnell, Janet A. *Supporting the Troops: The U.S. Army Corps of Engineers in the Persian Gulf War.* Alexandria, VA: Office of History, U.S. Army Corps of Engineers, 1996.

Nielsen, Richard. Research interview. February 28, 2004.

Ploger, Robert R., Maj. Gen. *Vietnam Studies: U.S. Army Engineers, 1965–1970.* Fort Belvoir, VA: Office of History, U.S. Army Corps of Engineers, 1974.

Potter, William E. *Engineer Memoirs: Major General William E. Potter, Retired.* US ACE, Office of the Chief of Engineers, Pub. 870-1-12, July 1983.

Stroupe, Wayne A. Research interview February 28, 2004 at Public Affairs Office, Airfields and Pavements Division (APD), Geotechnical Laboratory, US Army Engineer Research and Development Center.

Stubbs, Mary Lee and Stanley Russell Connor. *ARMOR-CAVALRY Part I: Regular Army and Army Reserve.* Washington, D.C.: Office of the Chief of Military History, United States Army, 1969.

Time. "A Laboratory Flies." July 7, 1941.

Time. "B-19." April 28, 1941.

Turhollow, Anthony F. "Airfields for Heavy Bombers." In Barry W. Fowle, ed. *Builders and Fighters: U.S. Army Engineers in WWII.* Pub. EP 870-1-42. Fort Belvoir, VA: Office of History, U.S. Army Corps of Engineers, 1992.

U.S. Army Corps of Engineers. *History of the Geotechnical Laboratory* (incomplete draft report). Vicksburg, MS: USACE Waterways Experiment Station, Geotechnical Laboratory, 2005.

Index

A

American Association of State Highway Officials, AASHO, 155
AASHO Road Test, 244
American Association of State Highway and Transportation Officials, AASHTO, 265, 277
Abbott, N. B., 99–102
Airfields
 recent conflicts, 200–201
 World War II and, 172–200
  construction abroad, 185–196
  demobilization, 196–197, 201
  heavy plane problems, 174–176
  runway research and development, 177–184, 198–199
Ahlvin, Richard, 200
Alcatraz Asphalt Company, 80–81
Andrews, Major General Frank, 176
Army Corps of Engineers. *See* Corps of Engineers
"Artificial" asphalt, 50–51, 71–95
 Bitulithic Macadam, 86–87
 delivering asphalt, 74–76
 making asphalt from crude oil, 80–81
 rural highways, 72–74
 success of, 93–95
 Warrens' asphalt dynasty, 81–85
Ash, H. W., 111
Asphalt
 airfields. *See* Airfields
 "artificial" asphalt. *See* "Artificial" asphalt
 avenues in Washington, D.C. and elsewhere. *See* Asphalting the avenues
 composition of, 15, 17
 environmental concerns and, 12, 257, 270–273, 281–282
 first 50 years. *See* Early advances in mixing and laying (first 50 years)
 future of, 282–283
 modern developments. *See* Modern developments
 in National Parks. *See* National Parks
 Route 66. *See* Route 66
 rubber, rubberized, 280–281
 superhighways. *See* Superhighways
 tycoon of (Amzi Lorenzo Barber), 56–69
Asphalt Association, 239
Asphalt Company of America, 64
Asphalting the avenues, 38–55
 accidentally asphalting macadam, 48
 "artificial" asphalt, 50–51
 competing pavement materials, 41–48

Constitution Avenue (Washington, D.C.), 14, 41
Fifth Avenue (New York), 50–51, 91
lake asphalt, 51–54
Massachusetts Avenue (Washington, D.C.), 14
Newark, New Jersey, 51–52
Pennsylvania Avenue (Washington, D.C.), 14, 38–41, 54–55
rock asphalt, 48–49
Asphalt Institute, 177, 179, 180, 199, 239, 240
Asphalt research, 267–277
Asphalt Trust, 63–65, 67–68, 76–78, 89
Asphalt tycoon (Amzi Lorenzo Barber), 56–69
Autobahns, 206–207
Aveling and Porter, 102
Avery, Cyrus Stevens, 154–156

B

Baker, Newton, 130
Barber, Amzi Lorenzo (the Asphalt Tycoon), 51, 56–69, 77, 80, 89–90, 107
Barber, Harry, 160
Barber Asphalt Paving Company, 58, 61–63, 71, 80, 85, 94–95, 97, 106–107, 109–110, 150–151
Barber-Greene Company, 160–162, 167, 205, 248, 251
Barber-Greene paving train, 160–162, 204–205
Barber-Greene floating screeds, 160–161, 252–253
Barford, William, 102
Barrett Company, 166
Bass, Ray, 274
Beagle, C. W. "Duke," 243
Bechtel, Stephen D., 237
Beck, David, 237
Belgium blocks, 43–44
Bell, Grady, 162
Bermudez Asphalt Paving Company, 64–66
Best Road and the Right Way to Make It, The, 69, 73
Bitulithic Macadam, 86–92, 107
Blanco, Guzman, 66
Boggs, Congressman Hale, 238
Brock, Don, 249–250
Brown, Ray, 274–275
Broder, David, 259
Buggyaut, 72
Bureau of Public Roads (BPR), 131–133, 135–137, 149, 206
Bush, President George Herbert Walker, 259
Butler tailgate spreader, 167
Byrd, Russell, 156

Acknowledgments

In January 2004, Margaret Cervarich of the National Asphalt Pavement Association called me and said, "We'd like you to do for asphalt what you've done for the Interstate System," referring to the book I'd written on that subject. I was hooked on the notion at first mention. Heartfelt thanks go to the entire membership and staff of the National Asphalt Pavement Association for commissioning this book and me to write it.

David Newcomb, who was extremely helpful while I was writing *The Roads That Built America*, had suggested that NAPA contact me to write this one, and for that I am grateful. He also was an important source of knowledge and constantly challenged me never to forget the main character of this book—asphalt.

Mike Acott, as the president of NAPA, provided direction and boundaries for the research and development of the book. Turning over the book's management to Margaret Cervarich, he put it in the hands of a skilled editor and project manger. Whenever I was in a jam, Margaret offered advice and working solutions.

Sarah Stillerman, NAPA's desktop publisher, provided expeditious support and her good sense of humor. My own deep regard and respect to each staff member of the NAPA. They have been wonderful to work with—always making me feel as if I were one of them.

Following are my best effort to express my gratitude to each person and organization that contributed knowledge, sources, and ideas toward this work. It is my hope that all who contributed to this book understand that I am indebted to them.

Jane Neighbors' contributions as editor have had an invaluable impact on the entire book. A mother of five, she once said that in raising children, the most important things are love, attention, and consistency. She's given all three to this effort. In addition, she researched and wrote part of the chapter on the National Parks.

Richard Berenson is a graphic genius. An illustrated book, such as this, is really two books in one—the written and the graphic. Richard marries text with pictures in a way that makes turning the pages exciting. Over the past five years he has become a friend, and as the proprietor of Berenson Books and Design, LLC, he has been an advisor to me on matters of publishing.

Peter Wilson of Barriere Construction Co. of New Orleans was the first NAPA member I interviewed for the book. Peter introduced me to the industry and his family's personal connection to it.

The Boston Athenaeum permitted me to photograph every page of John Loudon McAdam's book, *System of Road Making*, in order to thoroughly review the nearly 200-year-old publication.

Professor B. Holley, a member of the history department at Duke University since 1947, contributed mightily to the first half of the book through his paper "Blacktop: How Asphalt Paving Came to the Urban United States."

Todd Lynn of APAC, Inc., the successor firm to the Warren Brothers Company, personally saw to it that pictures of the Warren family and other treasures were converted from glass slides to reproducible digital images.

Don Nielsen, a former executive of Union Oil and a former chairman of the board of the Asphalt Institute, was a major consultant for chapter 6.

Mike Elliott, founder of Gencor Industries, Inc. helped bring to life some of the colorful personalities of asphalt paving. He ensured the accuracy of the information about asphalt machinery in chapter 7 and also contributed images.

Jim Roberts and Ross Kashiwagi of Granite Construction Co. suggested the idea of a chapter on paving in the National Parks. Together they tracked down images and stories of their firm's work on the roads of Yosemite Park and helped me meet George Wagner, a longtime member of Granite's family of employees. Speaking with George and reading his oral history was one of the highlights of my research.

Jim Powell, one of the most highly regarded historians on Route 66, and Ramona Lehman of Lebanon, Missouri, are cofounders of the new Route 66 Museum in Lebanon. They provided access to original maps and articles on the highway.

Jack, Don, Charlie, and Dan Gallagher kicked off my Route 66 trip from the offices of Gallagher Asphalt outside Chicago, Illinois, sharing their families' well-documented history in asphalt paving. Larry Lemon of Haskell Lemon Construction in Oklahoma, was also helpful in teaching me about paving the Main Street of America. Peter Kemmetern of Heckert Construction, in Pittsburg, Kansas shared his knowledge as I passed along the old traces of Route 66 near his offices. Herb Tiffany of Tiffany Construction discussed his father's paving of the Mother Road.

Chris Kirby of Barrett Paving Materials Inc. walked me through his company's famous Tarvia product and its connection with the some of the earliest paving jobs along America's rural highways. Tom Peterson of the Colorado Asphalt Association was helpful in discussing the use of Trinidad lake asphalt in modern times.

Alfred Chock, Jr., took on the research and writing of chapter 11, "Paving the Way to Victory." He spent untold hours away from family, while carrying his job's full-time workload, interviewing, sourcing, and writing a fascinating tale. Credit for that chapter's text and much of its imagery belongs to him.

Bruce D. Pelletier of the Maine Turnpike provided images of the early days of the first asphalt superhighway. Paul Yarossi and Barbara Pruitt of HNTB Companies shared graphic material on the Maine and New Jersey Turnpikes, which HNTB was instrumental in planning and constructing. Cliff Heath was involved in the construction of the New Jersey Turnpike and toured the nation in the 1980s promoting asphalt by using the New Jersey Turnpike as an example.

Dan Holt, director of the Eisenhower Library in Abilene, Kansas, and his staff, especially Herb Pankratz, gave me wonderful advice on sources for chapter 13.

Bart Mitchell guided me in exploring historical material for the chapter on the Interstates. Chuck Deahl of BOMAG Americas, who has been selling heavy equipment to the industry for decades, was an early advocate for the book, enthusiastically offering advice, tales, and sources. He connected me in a conference call with two of his colleagues, Tom Skinner of Skinner Consulting and Jack Farley, formerly of Barber-Greene, which contributed to the chapter "Ike's Grand Paving Plan." Tom has been instructing operators on how to use heavy asphalt equipment. Jack was involved in the manufacture of paving machines and products for Barber-Greene. He and his wife, Helen, researched photos for this book at the Aurora (Illinois) Historical Society.

Ray Brown, director of the National Center of Asphalt Technology's facility at Auburn University, put its scientific discoveries into lay terms. He and his staff, especially Buzz Powell and Dave Timm, Ph.D., were of special help, touring me around their laboratory and the test track.

Since the last chapter in this book covers recent events in the development of asphalt pavements, Margaret Cervarich wrote the entire chapter. I collaborated with her on the chapter's outline and integration with my own text, but beyond that and a review and critique of the final work, she and her colleagues produced the last word in this exciting history.

Judy Hornung, who manages the NAPA library, located old books for research. Melissa Nicienski, a Master of Library Science, provided consistent and immediate research material for this book.

In my five years of writing books on our nation's highways and byways, I learned much about the world of publishing from Barbara Morgan and her team of professionals, especially editor Marjorie Palmer, at Barnes & Noble.

I appreciate the advice given by Ronald White and his wife, Valerie, of Superior Paving Corp. that the book be an homage to the men and women involved in laying asphalt.

Jay Hansen of NAPA's government affairs department suggested significant subjects for research of asphalt history in the mid- to late-20th century. Chuck MacDonald, a journalist by training, emphasized that "a lively and clear voice is going to be crucial." Roger Sandberg, an expert on equipment and procedures, made an invaluable contribution with the final edits of imagery.

Dr. J. Don Brock turned over volumes of material and opened Astec Industries' treasure chest of graphics in order to make this book as visual as it is.

Thanks to Thomas W. Hill, president of Oldcastle Materials Inc., for his views of the evolving industry.

Richard and Carey Moore and their son-in-law Patrick Nelson gave insights into modern asphalt. Alan Scott and James Taylor of Aggregate Industries helped me in the specifics, art, and nuances of asphalt paving.

Thanks to Matthew Jeanneret and Bill Toohey for their counsel and encouragement to take on this project. And to Chris Nagle for his stimulating suggestions regarding artwork. Dana Thomas, who reports on arts and entertainment for *Newsweek,* was an unexpected source of ideas.

Ike Williams is a talent agent extraordinaire, and I am fortunate that he is mine. Over the last several years, I have benefited from his sage advice and solid directives. Hope Denekamp, his "right hand"of 15 years, is always positive and reassuring, listening to my concerns and questions and resolving them directly and quickly. Alexis Rizzuto, working with Ike and Hope, helped lay out a proposed story line. I am grateful to all of them.

A very special thanks to Professor Joe P. Mahoney of the University of Washington for his wisdom and wit in reviewing the manuscript and offering his story of asphalt.

Richard Weingroff of the Federal Highway Administration reviewed each page of the manuscript and challenged notions, offered lengthy feedback, and provided additional sources and imagery.

Scott Taylor, a college friend and an editor by profession, can uncover the slightest mistakes—just the individual for reviewing the final manu-script and making a more presentable product.

As an extra bonus for my work on highways, Scott Goldie, a producer for the History Channel's *Modern Marvels,* employed me to assist him in researching his documentary "Paving America."

I am dedicating this book to my cousin Danielle (Dana) Jennings, whom I first met when we were in our mid-twenties. We found that we were living and working within blocks of each other in Washington, D.C. Together we continue to peel back the mysteries of our family and its heritage. I can only hope that our legendary road-building and asphalt-paving great-grandfather, Sunny Jim, would be proud of our devotion to his legacy. Dana's closest friend, Greg Stevens, became my friend. With Dana's encouragement and Greg's support, I landed an appointment to the White House. Thanks to both Greg and Dana, my horizons were broadened and my journalistic confidence raised so that I am able to take on works such as this book. Sadly, Greg departed this world in February of 2005. Dana and I attempt to honor Greg's passion for living life to its fullest by emulating his restless desire to do it all.

Ji Jin is a beautiful and brilliant woman whom I am fortunate to call my wife. I owe her an unpayable debt for sacrifices made so that the manuscript for this work could be delivered on time. She often reminds me of how fortunate I am to have been asked to record such an important part of American history.

I sent every first draft and revised ones to my father, Dan McNichol.

Without his constant praise I would have been less able to face subsequent chapters. Both he and my mother, Ann McNichol, kept my morale high and provided encouragement. I love them dearly for all they have done and continue to do for me in my world of writing.

My extended family and friends are at the core of every important decision I face. Joe McNichol, my father's cousin, used his detective skills as an officer of the law to track down and meticulously log our family's history, especially Sunny Jim's. My friend Greg Wolfe was instrumental in the administration of this book. He has been and remains an inspiration and an example of how to conduct my own life, both professionally and personally. John Longley created an intellectual curiosity about literature when we were kids. His sister (I like to consider her mine too), Megan Longley, reviewed chapters and steered the narrative voice.

Thanks to Senator Jack Quinlan, for his influences as a lifelong mentor of writing, political storytelling, and character development.

My aunt Sheila and uncle Bob and my sister and brother-in-law Mega and Paul Scarpetta provided support on multiple fronts. My cousin John Peebles assisted me in arranging many of my long road trips, especially the 3000-plus-mile Route 66 trip.

I did much of the writing of this book at Starbucks No. 7377 on Atlantic Avenue in Boston. Ann-Marie, Chris, Delanie, Danie, Jenny, Kathy, Katie, Laura, Little-Laura, Scott, and Steve kept it from feeling like solitary confinement.

Dan McNichol
Boston, Massachusetts
October 2005

Acknowledgments from Alfred Chock, Jr., for Chapter 11

For the chapter "Paving the Way to Victory", I wish to acknowledge the contributions of Richard Ahlvin, P. E. He generously shared the technical knowledge and experience gained from his service as an army combat engineer in World War II and as a military pavement engineer and consultant for the past 60 years, and he reviewed several drafts of the chapter. He was recently honored by NAPA for his lifetime achievements.

Wayne Stroupe of the U.S. Army Corps of Engineers, Engineer Research and Development Center, provided historical documents, manuals, and photographs. In addition, the following people were instrumental in researching and preparing the chapter: Lara Tobias (Seabees Museum), Richard Mason, Walter "Mickey" Cereoli (NMCB1, 1970), Capt. Kelii H. Chock (USAF), Margaret Cervarich, Jane Neighbors, Dr. David Newcomb, and Dr. Joe Mahoney.

I would also like to thank Dan McNichol for the opportunity and the encouragement. Most of all, I am indebted to my wife, Bonnie Mason, who helped me sift through the history to find the real story.

Acknowledgments from the Publisher

Publishing the history of an industry would be a once-in-a-lifetime opportunity—and challenge—for anyone who makes her living handling communications for a trade association. I am grateful to the executive committee, board of directors, and members of NAPA for the challenge and the opportunity to produce this book. In particular, I thank the sponsors whose financial support has underwritten this enterprise so handsomely. Dick Stander and other members of the Council of Past Presidents and Chairmen were early champions of the idea. Bart Mitchell chaired the NAPA 50th Anniversary Task Force, which provided direction and oversight for the project. I fervently hope that all the members of NAPA will find that the book fulfills their vision.

Who would think of beginning a book about asphalt with a discussion of Roman roads? Dan McNichol! His enthusiasm, energy, and dogged determination to tell a "ripping tale of asphalt" in an engaging manner have carried this project from inception through conclusion.

I want to add my thanks to several people whom Dan has already acknowledged. Richard Weingroff at the Federal Highway Administration and Jim Powell, road historian, patiently reviewed text, offered advice, and supplied images to enhance the storytelling.

Richard Berenson of Berenson Design and Books created the look of the book, researched images, and was a gold mine of advice and information about the book publishing process. I relied heavily on Jane Neighbors, the editor, for all sorts of help. She not only made sure the grammar, punctuation, and spelling were consistent but also researched and drafted part of chapter 9, wrote captions for much of the book, and contributed in many other ways. Both Richard and Jane helped immeasurably. Also, many thanks to our indexer, Jerry Ralya.

Several distinguished scholars and industry experts reviewed the drafts of the chapters and gave valuable input. At the National Center for Asphalt Technology (NCAT), they were E. Ray Brown, Ph.D., director; Prithvi "Ken" Kandhal, Ph.D., associate director emeritus; and Raymond "Buzz" Powell, Ph.D., test track manager. At the Transportation Research Board, Neil Hawks, director of special programs, and Linda Mason, communications manager of the special programs division. Also, Frank Fee, CITGO Asphalt Refining; Professor Moreland Herrin of the University of Illinois Urbana-Champaign; Gerald Huber, Heritage Research Group; Professor Joe P. Mahoney, University of Washington; Charles F. Potts, Heritage Construction & Materials; and former University of Minnesota Professor Gene Skok, who is now Secretary-Treasurer for both the Association of Asphalt Paving Technologists and the International Society of Asphalt Pavements.

Butch Wlaschin at the Federal Highway Administration and Tim Davis, Ph.D., at the National Park Service provided much information about paving in the National Parks for chapter 9. Tonya Holloway of APAC, Inc., worked to get usable electronic files of very old glass negatives. At the New Jersey Turnpike Authority, Richard Raczynski helped verify the accuracy of much of chapter 12, and Joe Orlando supplied many images.

Each member of the staff of NAPA has contributed to this book. Mike Acott, the president, provided guidance, direction, and support. He made sure the finished book would be consistent with NAPA's members' vision for it. Tracie Christie, Úna Connolly, R. Gary Fore, Joanie Frykman, Jay Hansen, Kent Hansen, Diane Hindman, Howard Marks, Nancy Lawler, Chuck McDonald, Carol Metzger, Dave Newcomb, Roger Sandberg, Sarah Stillerman, Lauren Ward, Kim Williams, and Helen Yurevitch reviewed various drafts and provided input based on their areas of expertise. Dave Carskadon helped with computer matters. Sarah Stillerman also researched photographs, organized images, burned innumerable photo CDs, and provided assistance in any number of ways. Judy Hornung found recondite texts in the NAPA library and reviewed proofs. Ester Marchio reviewed proofs and braved the NAPA archives in order to assist with research. Colleen Vaniglio and Carolyn Wilson gave accounting support.

Finally, I wish to thank my husband, Frank, whose patience and support are never-ending, and my children, Frank III and Rachel, who put up with endless hours of proof-reading, fact-checking, e-mailing, and telephone calls while I should have been hanging out with them.

While this project has benefited from the support and input of many people, the responsibility for any errors, omissions, or faults is mine.

Margaret Blain Cervarich
Lanham, Maryland
October 2005

Picture Credits

Astor, Lenox and Tilden Foundations; 127 courtesy USDOT/FHWA; 128–129,130 top, 131 Dwight D. Eisenhower Library, Abilene, KS; 132 © Humanities and Social Sciences Library, Miriam and Ira D. Wallach Division of Art, Prints and Photographs, The New York Public Library, Astor, Lenox and Tilden Foundations; 133 courtesy USDOT/FHWA; 134–135 © Owaki-Kula/CORBIS; 136–137 bottom Library of Congress, Prints and Photographs Division; 137 top © Michael T. Sedam/CORBIS; 138 Library of Congress, Prints and Photographs Division; 139 Library of Congress, Prints and Photographs Division, The Bancroft Library, University of California, Berkeley; 140 courtesy American Association of State Highway and Transportation Officials (AASHTO); 141 courtesy George Wagner, Granite Construction Co.; 142–143 © ENA/Popperfoto, Retrofile; 144–145 courtesy of AASHTO, from the James R. Powell Collection; 146 James R. Powell Collection; 147 top © Bettman/CORBIS; Inset © Humanities and Social Sciences Library, Miriam and Ira D. Wallach Division of Art, Prints and Photographs, The New York Public Library, Astor, Lenox and Tilden Foundations; 149 top courtesy of AASHTO, from the James R. Powell Collection; bottom courtesy Ames Public Library; 152 collection of the author; 155 © PEMCO - Webster & Stevens Collection, Museum of History and Industry, Seattle/CORBIS; 156 collection of the author; 157 Smithsonian Institution, Gift of the Phillips Petroleum Company; 158 © Science, Industry and Business Library, The New York Public Library, Astor Lenox and Tilden Foundations; 161 courtesy Aurora Historical Society, Aurora, Illinois; inset courtesy Keith R. Schmidt, Caterpillar Inc.; 163 National Oceanic & Atmosphere Administration (NOAA); 164 © Bettman/CORBIS; 165 Will Rogers Memorial Museum, Claremore, Oklahoma, photo by Dan McNichol; 167 courtesy Aurora Historical Society; 169 © Bettman/CORBIS; 170–171 © Arthur Rothstein/CORBIS; 172–173 courtesy NAVFAC Historical Program, Seabees Museum; 175 courtesy Goleta Air & Space Museum (www.air-and-space.com), inset courtesy Hill Aerospace Museum, Hill Air Force Base, Utah; 179 Library of Congress, Prints and Photographs Division, photo by Alfred T. Palmer; 180, 181, 183 courtesy U.S. Army Corps of Engineers Waterways Experiment Station; 185 www.snapshotsofthepast.com; 187 courtesy U.S. Army; 188 Illustrated London News, courtesy University of San Diego History Department; 190 top left © CORBIS, top right and bottom © Time Life Pictures/Getty Images; 193 © CORBIS; 194 © Bettman/CORBIS; 195 © courtesy Raymond J. Schumacher Collection, www.rjs.org; 196 © CORBIS; 197 courtesy Corvair/Goleta Air & Space Museum (www.air-and-space.com); 198 courtesy Center for Southeast Louisiana Studies, Southeastern Louisiana University; 199 © Bettman/CORBIS; 200, 201 courtesy HQ, USACE, Office of History; 202–203 National archives; 205 courtesy Barber-Greene; 207 © Bettman/CORBIS, inset © Hulton-Deutsch Collections/CORBIS; 208, 209, 210 courtesy Maine Turnpike Authority; 213, 214, 215, 216–217, 218, 220–221 courtesy New Jersey Turnpike Authority; 222, 223, 224–225 top National Archives; 224 bottom, 225 bottom courtesy New Jersey Turnpike Authority; 227 Special Collections and University Archives; Rutgers University Libraries; Rutgers, The State University of New Jersey, inset © Roger Wood/CORBIS; 228, 231 © Bettmann/CORBIS; 232–233 © Raymond Gehman/CORBIS; 234 Library of Congress, Prints and

Photographs Division; 235, 236, 237, 238 top © Bettmann/CORBIS; 238 courtesy Sheldon W. Hayes, from Hayes family collection; 239 courtesy Cadillac Asphalt Paving Company; 241 courtesy USDOT/FHWA; 242 courtesy Maryland Transportation Authority; 243 courtesy U.S. Army Corps of Engineers Waterways Experiment Station; 244 courtesy USDOT/FHWA; 245 courtesy Gencor Industries, Inc.; 246 courtesy Wyoming Highway Department; 248 courtesy Astec Industries, Inc.; 252 courtesy Ingersoll-Rand Company; 253 courtesy Terex Roadbuilding; 257 courtesy Gallagher Asphalt Corporation; 263 large photo courtesy CMI Corporation; top right, middle, bottom courtesy MAK Milling; 266 courtesy Staker & Parson Companies; 267 courtesy GE Environmental Services; 268 courtesy National Center for Asphalt Technology (NCAT), Auburn University, Alabama; 269 courtesy Roadtec, Inc.; 270 top courtesy Richard Schreck, Virginia Asphalt Association, Inc.; 272 NAPA photos by Frank Cervarich; 274, 275 photos by Jim Killian, courtesy National Center for Asphalt Technology, Auburn University, Alabama; 276 top courtesy Arkansas State Highway and Transportation Department, bottom courtesy Antigo Construction Inc.; 278 courtesy Flexible Pavements of Ohio; 279 courtesy Asphalt Pavement Alliance; 280 large photo courtesy URS Corporation, inset NAPA photo by Dave Newcomb; 282–283 PhotodDisc/Fotosearch.

THE NATIONAL ASPHALT PAVEMENT ASSOCIATION
thanks the sponsors whose generosity
has made this book possible.

GOLD SPONSORS

Ajax Paving Industries, Inc.

Colas Inc.

The Cummins Construction Co.

Haskell Lemon Construction Co.

Lehman-Roberts Company

Superior Paving Corp./E. Stewart Mitchell, Inc.

Vecellio Group, Inc.

SILVER SPONSORS

BP Bitumen

Commercial Asphalt Co.

Gencor Industries, Inc.

Granite Construction Incorporated

Ingersoll-Rand Company

Maxam Equipment, Inc.

Terex

Wirtgen America